Metaverse and Tourism

This innovative and timely book presents an in-depth analysis of how the metaverse revolutionizes tourism management, transforms consumer behavior, and motivates tourists to visit destinations.

Adorned with illustrative tables, figures, and diagrams throughout, the volume is data-led and explores how metaverse experiences affect tourist satisfaction and loyalty toward metaverse tourism. It also takes a future-focused approach and looks at how the technologies of metaverse tourism will lead to a new level of immersive virtual reality. The book considers the metaverse in relation to the United Nations Sustainable Development Goals and post-COVID-19 society and offers practical insights on the topic.

This book will be of pivotal interest to students, scholars, and academics in the fields of tourism planning and policy, tourism economics, tourism behavior, and tourism development, as well as those with an interest in these areas more generally.

Marco Valeri is Associate Professor of Organizational Behavior at Niccolò Cusano University, Italy. His research areas include sustainability and green practices, strategy implementation, knowledge management, family business, crisis management, information technology, network analysis, and international hospitality management. He is a member of several editorial boards of international tourism journals and is a reviewer and editor of several handbooks on entrepreneurship, tourism, and hospitality management.

Ahmad Albattat is Associate Professor in the Graduate School of Management, Post Graduate Centre, Management and Science University, Malaysia. His research areas include hospitality management, hotel, tourism, events, emergency planning, disaster management, sustainable development goals, and human resources. He is an active member and on the Editorial Board of several scientific international journals, as well as a reviewer, an author, and an editor for book projects.

Routledge Insights in Tourism Series

Series Editor: Anukrati Sharma
Head & Associate Professor of the Department of Commerce and Management at the University of Kota, India

This series provides a forum for cutting edge insights into the latest developments in tourism research. It offers high quality monographs and edited collections that develop tourism analysis at both theoretical and empirical levels.

Pseudo-Authenticity and Tourism
Preservation, Miniaturization, and Replication
Jesse Owen Hearns-Branaman and Lihua Chen

Developing Industrial and Mining Heritage Sites
Lavrion Technology and Cultural Park, Greece
Taşkın Deniz Yıldız

Innovation Strategies and Organizational Culture in Tourism
Concepts and Case Studies on Knowledge Sharing
Edited by Marco Valeri

Tourism and Poverty Alleviation in Nature Conservation Areas
A Comparative Study Between Japan and Vietnam
Nguyen Van Hoang

Sport Tourism, Events and Sustainable Development Goals
An Emerging Foundation
Edited by Anukrati Sharma, Miha Lesjak and Dusan Borovcanin

Tourism, Philanthropy and School Tours in Zimbabwe
Problematising "Win-Win" Discourses
Kathleen Smithers

Metaverse and Tourism
Rethinking Implications on Virtual Reality
Edited by Marco Valeri and Ahmad Albattat

For more information about this series, please visit: www.routledge.com/Routledge-Insights-in-Tourism-Series/book-series/RITS

Metaverse and Tourism

Rethinking Implications on Virtual Reality

Edited by Marco Valeri and Ahmad Albattat

Routledge
Taylor & Francis Group

LONDON AND NEW YORK

First published 2025
by Routledge
4 Park Square, Milton Park, Abingdon, Oxon OX14 4RN

and by Routledge
605 Third Avenue, New York, NY 10158

Routledge is an imprint of the Taylor & Francis Group, an informa business

British Library Cataloguing-in-Publication Data
A catalogue record for this book is available from the British Library

ISBN: 978-1-032-80470-5 (hbk)
ISBN: 978-1-032-80471-2 (pbk)
ISBN: 978-1-003-49700-4 (ebk)

DOI: 10.4324/9781003497004

Typeset in Times New Roman
by Apex CoVantage, LLC

Book published with economic contribution of the Department of Political, Legal and Sociological Sciences, Niccolò Cusano University

Contents

Tables

Figures

Contributors

Manpreet Arora is Senior Assistant Professor of Management at the Central University of Himachal Pradesh, Dharamshala, India. Her research interests encompass accounting, finance, strategic management, entrepreneurship, qualitative research, and microfinance.

Pınar Başar is Associate Professor at Faculty of Business Administration, Istanbul Commerce University, with expertise in business administration, specifically management, and organization. She teaches various courses on business, management, organization, entrepreneurship, strategic management, leadership, innovation marketing, and market research.

Isha Kumari Bhatt is Assistant Professor at Daulat Ram College, University of Delhi, pursuing her Ph.D. degree at the Delhi School of Economics, University of Delhi. Isha's research interests revolve around cutting-edge topics in tourism and marketing, including smart tourism, exploring innovative approaches to enhance the tourist experience through technology, and green tourism.

Martha Isabel Bojórquez Zapata is Doctorate in Administration from the Universidad del Sur. She is a full professor of subjects such as investment projects, budgets, and working capital management at the Facultad de Contaduría y Administración de la Universidad Autónoma de Yucatán. Her research areas include competitiveness in small- and medium-sized enterprises (SMEs), strategies, shared value, and lean philosophy. She is a member of several editorial boards of international journals that publish business topics.

Mohammed Reda Boukhecha is a fresh laureate in Business Administration from the University of Padova Department of Economics and Business Sciences "Marco Fanno," Italy. His areas of expertise include tourism, sustainable development, international coordination, and trading.

Adriana Otálora Buítrago is Doctorate in Political Studies from the Universidad Externado de Colombia, where she is the leader of citizenship, ethics, and politics research and a full professor of subjects such as human, integral, and sustainable development. Her research areas include citizenship studies, citizenship education, public policy, democracy, political philosophy, and human development.

Nishita Chatradhi is a researcher and has co-authored multiple papers, edited numerous research documents, and facilitated government-funded projects. Her expertise lies in academic writing, with a thematic focus on livelihood development, sustainability, and entrepreneurship.

Seyed Ilia Daneshpour is a Ph.D. researcher in design and conducts research in the user experience realm in environments, like the metaverse, based on his previous experience developing digital and consumer products to improve human–computer interaction and the resulting experiences.

Nisha Devi is Assistant Professor in the Department of Commerce, Shri Ram College of Commerce, University of Delhi. Her research interests include human resource management, OB, entrepreneurship, and tourism. Currently, she is working on religious tourism, rural tourism, and sustainable development.

Fatima Zahra Fakir is Assistant Professor at the University of Padua, affiliated with the Department of Economics and Business Sciences "Marco Fanno." Her area of expertise lies in smart tourism, cultural heritage, and sustainable development.

Evangelia Kasimati is Senior Economist at the Economic Analysis and Research Department of the Central Bank of Greece. Her research and teaching interests lie in the areas of financial economics, inflation, tourism and sports economics, and macroeconomics.

Blandina Kisawike is Senior Lecturer in Marketing and Acting Deputy Vice Chancellor for Planning, Finance, and Administration at the University of Iringa. She has 15 years of teaching experience in various courses such as marketing, entrepreneurship, management, international business, and research methods.

Suneel Kumar is Professor in the Department of Commerce at Shaheed Bhagat Singh College, University of Delhi. He led a major ICSSR, IMPRESS, Ministry of Education, Government of India research project titled "Sustainable Rural Tourism: An Approach for Transforming Rural India – A Case Study of Himachal Pradesh." His research focuses on Tourism and Hospitality Management, Rural Tourism, Sustainability and Green Practices, Destination Branding, and Marketing. Currently, he is Co-Principal Investigator for the ICSSR project "Eco-Tourism: A Panacea for Sustainable Development in Himachal Pradesh." Dr. Kumar has presented about 50 papers at conferences and published over 35 research papers and book chapters.

Varinder Kumar is Assistant Professor in the Department of Commerce at Zakir Husain Delhi College (E), which is affiliated with the prestigious University of Delhi (India). His research primarily centers on the dynamic field of tourism, encompassing areas such as tourism marketing, sustainable tourism, community-based tourism, smart tourism, sports tourism, and rural tourism.

Małgorzata Kurleto is Professor at the Faculty of Management and Social Communication of the Jagiellonian University. In the scope of her research interests, Malgorzata deals with civil law, intellectual property protection, tourism law, risk management in tourism, protection of the natural environment, and the use of modern technologies in tourism.

Nasser Koleini Mamaghani is Associate Professor of Industrial Design at the Iran University of Science and Technology. His research interests include Kansei engineering, human-oriented design, universal design, ergonomics, physiological anthropology, and design methodology (QFD, Kano model).

Igor Mavrin is Assistant Professor at the Academy of Arts and Culture in Osijek (Republic of Croatia). His research fields include European Capitals of Culture, cultural management, urban economics, cultural and creative tourism, virtual tourism, and creative industries.

Mukesh Kumar Meena is currently Assistant Professor in the Department of Commerce at Delhi College of Arts and Commerce, University of Delhi, and holds a doctorate from the Department of Commerce, Faculty of Commerce and Business, Delhi School of Economics, University of Delhi. Dr. Meena's research and publication interests span diverse areas, encompassing human resources accounting, corporate governance, finance, corporate finance, accounting, and education.

Seyed Hashem Mosaddad is interested in design, emotion, technology evolution, and product design and has authored numerous articles on these topics. With an extensive background in teaching industrial design at various Iranian universities, he is currently Director of the Industrial Design Department at the Faculty of Architecture and Urban Planning at Iran University of Science and Technology.

Hassan Sadeghi Naeini has taught Ergonomic Design courses at various universities, including the University of Tehran (Iran), IUST (Iran), Poly-Techniques University (Iran), UPM (Malaysia), and as Visiting Researcher at KTH (Stockholm). He has substantial experience as Industrial Project Manager in Ergonomics, having worked in over 15 industrial sectors.

Alberto Gabriel Ndekwa is Senior Lecturer and Director of Research, Publication, and Consultancy at Ruaha Catholic University in Tanzania. Dr. Ndekwa has experience in business, information and communication technologies, and community development.

Sabri Öz is Associate Professor and Head of the Industrial Policy and Technology Management (SAPTEY) Department at Istanbul Commerce University. Her research interests include distribution, disadvantaged social groups, economic integration, logistics, technology, and digital transformation.

Manisha Paliwal is an accomplished academic and researcher with more than two decades of experience. She is currently serving as Experienced Professor and

Head of the Entrepreneurship Center and Deputy Head of the Research and Development Cell at Sri Balaji University, Pune. Her research areas include international business, strategic management, foreign exchange, and business ethics.

Antonio Emmanuel Pérez Brito is Doctorate in Strategic Planning and Technology Management from the Universidad Popular Autónoma del Estado de Puebla. His research areas include competitiveness in SMEs, strategies, shared value, and lean philosophy. He is a member of several editorial boards of international journals.

Athina Rentifi is currently studying toward an M.Sc. degree at the University of Athens in Geopolitical Analysis, Geostrategic Synthesis, and Defense and International Security Studies. She is working as Senior Economist at the Economic Analysis and Research Department of the Bank of Greece in the Balance of Payments Analysis Section.

Silvia Rita Sedita is Full Professor in Management at the University of Padova, where she teaches advanced marketing, sustainable business models, and economics and management of innovation. Her research agenda includes issues within the field of management of creativity and innovation in interorganizational networks.

Corina Tursie is Ph.D. Senior Lecturer at West University of Timisoara, Faculty of Political Sciences, Philosophy and Communication Sciences. With a Ph.D. degree in Political Sciences, she teaches EU public policies and policy analysis. She is a member of the CCECUT network, investigating the various public policies developed by cities in order to reach the goals of the European Capital of Culture program.

Nikolaos Vagionis is Scientific Researcher at the Center for Planning and Economic Research, Senior Researcher, and Deputy Scientific Director. He has also taught at the Hellenic Open University (2007 et seq.), at the M.Sc. degree in Tourism Business Administration, and at the Undergraduate Course of Tourism Management.

Preface

Metaverse notion relates to the building of virtual environments that are centered on social relationships. Metaverse is defined as the convergence of physical and digital universes, where users can seamlessly traverse between them for working, education and training, health, exploring interests, and socializing with others. Virtual reality experiences, in which viewers are put in a digitally changed environment, augmented reality overlays of the actual world, and even video games are all examples (Buhalis, 2023). Frequently, the notion of the metaverse is also related to users having control over a digital avatar, which is then utilized to converse with other metaverse participants. In addition to its social component, companies are investigating the potential the metaverse offers to contact consumers wherever they may be. Consequently, the metaverse has become one of the growing tourist trends. As metaverse tourism prospects continue to develop and user acceptance rises, the metaverse is anticipated to play a bigger role in the tourist sector, increasing communication and the customer or visitor experience. When investigating the concept of metaverse tourism, it might be important to have a thorough understanding of the tourist sector. This sector of the service business focuses on operations associated with the temporary relocation of individuals. It consists of transportation, lodging, entertainment, and food and beverage. When describing the concept of the metaverse, several individuals first fail to comprehend how it differs from existing virtual reality. In essence, the distinguishing characteristic of the metaverse is the construction of an interactive virtual environment; hence, virtual reality plays a part but is not required. Video games, interactive video content, augmented reality, and other related technologies, for instance, may be used to construct and explore interactive virtual worlds. This gap may be comprehended by considering virtual reality as only one of many conceivable technologies that might support metaverse tourist services.

The second concept that must be considered in understanding the metaverse and tourism is the notion that the metaverse has existed for longer than most people realize, but it has become a more well-known concept only lately as its technology has improved. There is a direct relationship between the growth of tourism and technology. Indeed, from the automated booking offices of the 1970s to the domestication of the Internet in the late 1990s, technology has always been employed in tourism to generate new patterns. The metaverse is a component of this development of the Internet, which employs more immersive technology to provide physical experiences – that is, experiences that blur the line between the real and the virtual. When it comes to tourism, Asia is a leader, with plans such as the Seoul Metaverse project, which wants to become the first major city in the world to join the metaverse, with a tourist circuit recreating the city's important sights. However, it is in France that we discover one of the most successful experiments with MoyaLand, a virtual tourism realm containing a tourist office, museums, an airport, and a historical center where people and visitors can travel around virtually through their avatars. This has produced several metaverse tourist prospects, which various nations and corporations are eager to exploit. The relationship between the metaverse and tourism is clear in the capability for producing virtual real-world items and places. As a consequence of a quantum leap in technological development, the barrier between what is physical and what is digital is becoming blurrier. The metaverse and its implications on a variety of sectors and our way of life have lately garnered considerable attention. The most recent advancements in metaverse technology have motivated numerous businesses to utilize this sophisticated tool, including the tourist industry.

The recent epidemic not only momentarily impacted the tourist sector but also revealed the industry's fragility. As more natural disasters and uncertainties are anticipated as a result of global warming and other causes, there is a good likelihood that individuals will choose for domestic travel over foreign travel. These deficiencies illustrate some of the most major ways in which the concept of metaverse for tourism is impacting the travel industry. Immersive virtual worlds could enhance customers' interactions with locations and in some circumstances even replace real travel without adversely affecting the business. In the future years, the metaverse has the potential to provide some of the most immersive travel experiences. Before making a purchase, the majority of buyers often test the goods. By permitting virtual experiences prior to vacation booking, the metaverse has the potential to completely transform the tourist business. This will promote the "try before you purchase" concept. The metaverse is seen as a catalyst rather than a replacement for current technology. Consequently, businesses such as hotel chains, tour firms, and travel agencies may advertise their services through the metaverse. Consequently, the metaverse can affect travel patterns. As additional metaverse travel and tourism choices emerge and user acceptance rises, it is expected that the metaverse will have a stronger influence on the tourist sector, boosting user experience and communication.

Structure of the Book

The book *Metaverse and Tourism. Rethinking implications on virtual reality* is the result of reflections involving research studies of different nationalities. The book contains 13 chapters written by 30 authors located in 13 different countries (Colombia, Germany, Greece, India, Italy, London, Mexico, New Zealand, Poland, Republic of Croatia, Romania, Tanzania, and Turkey) and affiliated with 19 different universities. The book will be an opportunity to stimulate academic research, approaches, and methods on how new technologies of metaverse tourism will lead to a new level of immersive virtual reality.

The book is structured in two parts. The first part focuses on *operations strategies in metaverse*. This part collects chapters that analyze how metaverse revolutionizes tourism management and how it transforms consumer behavior, motivating to visit destinations physically. The second part focuses on *experiences in metaverse*. This part collects chapters that analyze how metaverse experiences affect tourists' satisfaction and loyalty toward metaverse tourism.

The first chapter titled "Role of Metaverse in Future of Tourism, Hospitality, and Events: Insights From Bibliometric Review" is written by Mukesh Kumar Meena, Suneel Kumar, and Marco Valeri. The aim of this chapter is to the metaverse, which refers to a virtual reality space where users can interact with a computer-generated environment and other users in real time. It is a concept that has gained significant attention in recent years, especially with the development of virtual and augmented reality technologies. The metaverse is a virtual reality-based space where users can interact with a computer-generated environment and other users in real time. It is often described as a fully immersive and interactive virtual world. The concept gained significant attention due to its potential to revolutionize various industries, including tourism. In the context of tourism, the metaverse can offer new and exciting opportunities. It could enable virtual travel experiences, where users can explore destinations, landmarks, and cultural sites from the comfort of their homes. This could be particularly appealing to individuals who are unable to travel physically, whether due to financial limitations, health concerns, or other constraints. Furthermore, the metaverse could enhance the pre-travel planning phase by providing immersive and realistic simulations. Travelers could virtually explore hotels, resorts, and attractions to make more informed decisions about their travel arrangements. This virtual experience could help manage travelers' expectations and improve overall satisfaction with their chosen destinations. The findings of this study provide valuable insights into the opportunities and potential future scenarios that the metaverse offers for the hospitality sector. The concept of the metaverse refers to a virtual reality space where users can interact with a computer-generated environment and other users in real time. It is often described as a fully immersive and interconnected virtual world that goes beyond the traditional boundaries of the Internet. The metaverse has gained significant attention in recent years due to advancements in virtual reality, augmented reality, and other immersive technologies. About the future of hospitality, the metaverse presents intriguing possibilities for the industry.

The second chapter titled "Critical Aspects of Studies Relating to Virtual Tourism in the Context of Values Resulting From Sustainable Tourism, Nature Tourism, Ecotourism, and Cultural Tourism" is written by Małgorzata Kurleto. The chapter aims to analyze critical scientific papers related to virtual tourism in specific types of tourism. Additionally, it explores the potential relationship between tourism and the metaverse. The analysis is grounded in critical aspects of selected literature concerning virtual tourism in specific types of tourism. The study encompasses 36 thematically relevant literature items and an additional examination of 14 papers on selected types of tourism to underscore the value of real tourism. The analysis includes theoretical dissertations, secondary research, blogs, and guides. The analysis reveals a growing connection with the metaverse in the tourism sector, prominently manifesting in the emergence of virtual tourism. The literature study raises concerns about the application of virtual tourism in certain tourism types, particularly in natural or cultural tourism. This signals skepticism toward the preservation of crucial traditional values in real tourism and the potential harm this metaverse form may pose to "virtual tourists." The research substantiates the hypothesis that certain traditional tourism types possess values that virtual tourism cannot substitute. The study specifically refers to sustainable tourism, nature tourism, ecotourism, and cultural tourism, urging cautious consideration and limitations on the use of virtual tourism. It also underscores the significance of attending to the physical and mental health of individuals engaging in virtual tourism.

The third chapter titled "The Future for Business Competitiveness in the Tourism Sector: Tourism Agenda 2030 Perspective," is prepared by Antonio Emmanuel Pérez Brito, Martha Isabel Bojórquez Zapata, Adriana Otálora Buítrago and Marco Valeri. The aim of this chapter is to ascertain the novel strategies that improve/maintain competitiveness in the context of sustainability and the Sustainable Development Goals (SDGs) within tourism enterprises. Despite the abundance of scholarly literature examining the concept of sustainable competitive tourism, the discourse surrounding the interplay between tourism, competitiveness, and sustainability remains understudied. Furthermore, the industry has the formidable task of addressing the 2030 Agenda, as it has thus far failed to effectively mitigate environmental degradation and rectify social inequalities. This work employed a detailed review to offer a comprehensive understanding of the subject, recognizing actions focused on sustainable tourism that converge to increase the competitiveness of this important and strategic economic sector. Competitiveness in the tourism sector still maintains the economic approach that gave rise to the concept, and it is important to keep in mind the factors that have an impact on said competitiveness. Factors related to strategic management, finance, innovation, and marketing, among others, are also important. The most recent research topics are shifting toward heterodox approaches of economics such as ecological economics. Among sustainability issues, other global concerns such as climate change, financial sustainability, sharing economy, and sustainable supply chain management appear. Tourism appears as a strong link between these emerging topics and concerns and the nodes of competitiveness and sustainability. Meanwhile, competitiveness as a field of study has been displaced by sustainability, based on the migration to different approaches in economics and business studies.

The fourth chapter titled "The New Economic Paradigm Formed by Metaverse in the Tourism Sector: Touristic GIG Economy" is written by Sabri Öz and Pınar Başar. This chapter investigates how the use of metaverse platforms, which emerged with augmented reality (AR) and virtual reality (VR) technologies in the tourism sector, as a tool contributes to the GIG Economy. Chapter 4 examines the use of metaverse platforms in the tourism sector, the conveniences they provide, and their futuristic approaches. Academic trends were tried to be understood by conducting a literature review and bibliometric analysis in the same field. Afterward, PESTLE components were weighted as Political, Economic, Social, Technological, Legal-Regulatory, and Environmental dimensions, analysis, structured in-depth interviews, and analytical hierarchical process (AHP) analysis. As a result of the analysis, it was stated how important the economic factor is. The limitations of the analysis are also listed. The general criticism of AHP analysis also applies to this study, because the phenomenological study is based on subjective evaluations. However, opinions are consistent. On the other hand, the fact that the study was conducted with only 9 people with at least 15 years of experience in the tourism sector is among the limitations of the study. Research can be developed with more people. On the other hand, it is expressed as a separate limitation that the participants in the evaluation may think differently due to the definition and understanding of the GIG economy.

In the fifth chapter titled "Transforming Tourist Consumer Experience Through Virtualization: Does Metaverse Matter?" Alberto Gabriel Ndekwa and Blandina Kisawike analyze how the virtualization of tourism marketing practices transforms the tourist consumer experience in the metaverse age. A quantitative approach was used to collect quantitative data on how virtualization through metaverse transforms the tourist consumer experience. A standardized questionnaire was used to collect standardized and structured data for hypothesis testing. The targeted population was tourist consumers in Tanzania whereby a simple sampling technique was used to pick a sample from the targeted population. Results derived from a sample of 287 respondents show that virtualization through metaverse transforms the tourist consumer experience. The significant influence of the virtualization metaverse on the tourist consumer experience is explained by the fact that metaverse tends to offer virtual tour, information guide, customization, and blended virtual and physical worlds which enhance the tourist experience. This study added value to the body of literature by evidencing the link of the influence of virtualization through metaverse on transforming the tourist consumer experience. While this study was done in Tanzania, it is suggested that future studies be undertaken in other countries of Africa to provide further evidence of the influence of virtualization of tourism marketing practices on transforming the tourist consumer experience.

In the sixth chapter titled "Metaverse as a Socio-Technical Phenomenon in Hospitality and Tourism: An Approach to Sustainable Development Goals," authors Seyed Ilia Daneshpour, Nasser Koleini Mamaghani, Hassan Sadeghi Naeini, and Seyed Hashem Mosaddad aim to construct a framework and foundational components for a white paper that explores the transition from traditional electronic, smart, or virtual tourism to emerging metaverse-driven tourism. The focus is on aligning this shift with the SDGs, particularly emphasizing the eighth goal of "Decent work & economic growth." The authors review scholarly studies to examine the convergence of SDGs, the Metaverse, and designed tourism platforms, defining minimum requirements and standards similar to WEB2.0 development with an emphasis on sustainable development goals. Metaverse-driven tourism affects various aspects, categorizing users as suppliers and tourists. Suppliers benefit from the Metaverse but face complexity, with the quality of experience (QoE) as a crucial aspect. The authors introduce various rating tests and metrics to assess QoE, emphasizing its significance in crafting feature sets. Tourism managers, as suppliers, are encouraged to leverage the metaverse for tasks such as product creation, strategic decision-making, and pursuing sustainable goals. Additionally, the chapter highlights the importance of addressing the unique needs of older users in the advancing technological landscape, suggesting tailored learning solutions within the metaverse for this demographic. Consideration for factors such as emotional well-being, tech comfort, training effectiveness, and cognitive skills is crucial in this design effort. The authors stress the need to pay attention to the demographics of Generation Z and Alpha, aligning with their specific needs and requirements.

The seventh chapter titled "What Lies in the Current Scholarly Art of Work on Metaverse and Tourism? Exploring Future Research Directions" is written by Manpreet Arora and Marco Valeri. The aim of this chapter is to investigate how the megatrends that already shape the changes in the global and of course Greek tourism influence tourism and relevant income in Greece. Specifically, factors such as the need for digitalization and technological upgrade and transformation, climate change and the need for environmental protection and planning for sustainability, the emerging socio-demographic changes and planning for overtourism, security, quality, and hygienic/epidemic issues together with the reality of sharing economy and the emerging secondary destinations are addressed. In light of the above, the particularities and varying potential of Greek regions are examined. The methodological approach is based on a qualitative analysis, referring to secondary data and research. Our findings contribute to the tourism literature by providing valuable insights into a limited body of relevant academic research for Greece.

The eighth chapter titled "Tourism Agenda 2030 in Greece" is written by Evangelia Kasimati, Athina Rentifi, and Nikolaos Vagionis. The chapter examines the tourism agenda of Greece toward 2030, focusing on the new fresh vision for Greek tourism and targeting the increase of visitors, receipts, and average spending per trip, as well as the qualitative advancement of the product and the experience. In addition, this chapter presents action plans to further develop Greek tourist offerings and products in terms of destinations and clusters across the country. The methodological approach is based on a qualitative analysis, considering secondary data and research. Our findings contribute to the tourism literature by providing valuable insights into the tourism agenda toward 2030, especially since the body of such academic research for Greece is limited.

The ninth chapter titled "Journeying Beyond Reality: Exploring India's Metaverse Marvel" is written by Manisha Paliwal, Nishita Chatradhi, and Marco Valeri. This chapter aims to explore the convergence of the metaverse with India's thriving tourism industry and investigates the potential for unique metaverse experiences to enhance tourism. The tourism industry in India is rapidly growing and holds significant economic importance. India has been progressing in its competitiveness in the travel and tourism sector. The country boasts approximately 38 United Nations Educational, Scientific and Cultural Organization World Heritage sites, attracting tourists from around the globe. Integrating VR technology in tourism is transforming how people experience travel. The Ministry of Tourism, Government of India, has partnered with several travel technology startups to enhance the Incredible India promotional initiatives by offering virtual travel experiences. However, VR technology in tourism faces challenges, including cost, awareness, and technology limitations. Despite these challenges, there are significant opportunities for the tourism and hospitality sectors in India to leverage VR technology. Virtual tours can provide immersive experiences of destinations and help tourists make informed choices. The potential integration of the metaverse into the tourism sector could offer unique virtual experiences of landmarks and cities in India. The metaverse is a burgeoning digital frontier with the potential to revolutionize various industries, including tourism. India's tech industry and startups are well-positioned to contribute to the metaverse's development. However, these challenges must be addressed, such as digital inclusivity, regulation, and societal impacts. India's integration with the metaverse could have significant social and cultural implications, including fostering cultural unity and digital inclusion. Integrating VR and the metaverse in India's tourism industry presents opportunities and challenges. By addressing these challenges and leveraging the potential of VR and the metaverse, India can offer immersive and engaging experiences to travelers, enhancing its tourism sector and contributing to its digital evolution.

The tenth chapter titled "Digital Marketing Metaverse for Restoration of Tourism Performance After COVID-19 Business Crisis in Sub-Saharan African Countries: Experience From Tanzania" is written by Alberto Gabriel Ndekwa. This study aims to analyze the contribution of the post-COVID-19 digital marketing metaverse in restoring the performance of the tourism sector after the COVID-19 threat altogether in an effort to redress the Business Crisis experienced in sub-Saharan countries. The study deployed a quantitative and qualitative approach. The quantitative approach and matching statistical analysis enabled the assessment of the contribution of digital marketing metaverse in promoting the performance of tourism performance in Tanzania. On a similar note, a qualitative approach was applied in exploring how digital marketing metaverse can better restore the performance of the tourism sector after the threat of COVID-19. A systematic sampling technique was applied in picking a sample of 244 from the targeted population. Purposive sampling was used in selecting respondents for data collection. Partial least square structural equation modeling deploying Smart PLS 4 was called to bear in analyzing the quantitative data while thematic analysis was applied in analyzing qualitative data. The findings revealed a significant contribution of digital marketing metaverse to the restored performance of the tourism sector in sub-Saharan countries. The finding revealed a significant contribution of digital marketing metaverse based on digital information, digital knowledge sharing, and digital promotion initiatives. It is concluded that digital marketing metaverse significantly contributes to the restoration of the performance of the tourism market sector.

The 11th chapter titled "Virtual Tourism as the Sustainable Future of Travel" is written by Igor Mavrin and Corina Tursie. The chapter aims to detect the key trends of immersive reality technologies used in shaping the new tourism trends, with special emphasis on virtual tourism, based on VR technology, as the substitute for physical travel. The COVID-19 pandemic at the beginning of 2020 showed that virtual experiences could be a temporary supplement for touristic experiences when travel is not an option. Global policies and consumer trends immersed into sustainability, combined with travel risks, such as health issues and terrorism, together with longevity trends, add additional arguments for virtual tourism as the future of travel. Contemporary travelers are heavily relying on technology, including not only online and mobile platforms but also social media, searching for travel inspiration and recommendations. The concepts of immersive tourism and virtual tourism have been emerging in the past 20 years, allowing travelers to explore destinations, either in situ, enhanced by technology, or distantly, without leaving their homes or places of living. Scientific interest has also emerged in the same period. In the first decade of the 21st century, researchers have mainly been focusing on three-dimensional (3D) models and VR, while the focus from 2011 to 2020 has shifted to AR and other emerging immersive technologies. Although virtual tourism could be seen as a threat to the travel industry, it could also be monitored as a new niche of tourism, turning non-travelers into (virtual) travelers, and allowing travelers to extend their traveling experiences, generating new tourism markets.

The 12th chapter titled "Exploring Immersive Tourism: Insights From Padova's Tourism Managers" is written by Mohammed Reda Boukhecha, Fatima Zahra Fakir, and Silvia Rita Sedita. The study examines the impact of immersive digital technologies and digital content on tourists' experiences and engagement at cultural heritage sites in Padova, within the context of a smart cultural destination. Cultural heritage refers to the tangible and intangible elements inherited from the past that hold cultural significance and contribute to a sense of identity and continuity within a community or society. Smart tourism encompasses the integration of technology, data analytics, and innovative strategies to enhance the efficiency, sustainability, and visitor experience within the tourism sector. Immersive experiences involve the use of digital technologies, such as AR and VR, to create interactive and multi-sensory environments that fully engage and immerse participants in a simulated or augmented reality. The chapter seeks to understand how these concepts intersect and influence tourists' perceptions and decision-making processes. Specifically, the objectives include investigating the role of immersive digital technologies in enhancing tourists' experiences, assessing the influence of digital content on tourists' engagement with cultural artifacts, understanding tourists' preferences regarding the utilization of digital technologies, examining the integration of digital technologies into the tourism industry in Padova, and identifying strategies for offering immersive experiences that align with tourists' preferences and enhance their overall satisfaction and engagement. On the other hand, this study delves into the perspectives of key stakeholders in the tourism industry, including tourism managers, tourist guides, and relevant organizations. Through their insights, we aim to comprehensively examine the dynamic interplay between immersive experiences and the roles played by these industry professionals in shaping the tourist landscape. Through addressing these objectives, the chapter aims to contribute to the understanding of how immersive digital technologies can enhance cultural heritage tourism and inform future destination management and marketing strategies.

The 13th chapter titled "Impact of the Metaverse on the Tourism and Hospitality Industry" is written by Suneel Kumar, Varinder Kumar, Nisha Devi, and Isha Kumari Bhatt. The chapter investigates the transformative potential of the metaverse in revolutionizing customer experiences and business operations within the tourism and hospitality sectors. By incorporating AR and VR technologies, the metaverse bridges the gap between the physical and virtual realms, providing immersive and interactive experiences that enhance customer engagement and personalization. The chapter highlights the advantages of metaverse adoption, such as fostering travel enthusiasm, mitigating uncertainty for travelers, and creating creative marketing opportunities. Responsible incorporation of Metaverse technologies is essential to realizing their full potential. The research emphasizes the need for further examination of customer co-creation dynamics, the cognitive and emotional impacts of AR and VR, economic obstacles, and health and safety considerations. Ultimately, by addressing these challenges and capitalizing on metaverse opportunities, the tourism and hospitality sector can generate bookings, revenue, and industry innovation.

Part I
Operations Strategies in Metaverse

1 Role of Metaverse in Future of Tourism, Hospitality, and Events

Insights From Bibliometric Review

*Mukesh Kumar Meena, Suneel Kumar,
and Marco Valeri*

Introduction

The metaverse refers to a virtual reality space where users can interact with a computer-generated environment and other users in real-time. It is a concept that has gained significant attention in recent years, especially with the development of virtual and augmented reality technologies. In the context of tourism, the metaverse has the potential to revolutionise the way people experience and engage with destinations.

It is important to note that the metaverse is still an evolving concept, and its full potential in the future of tourism is yet to be realized. The actual implementation and impact of the metaverse in tourism will depend on technological advancements, user adoption, and industry innovation (Valeri, 2024b; Sinha et al., 2024). The metaverse is a virtual reality-based space where users can interact with a computer-generated environment and other users in real time. It is often described as a fully immersive and interactive virtual world. The concept gained significant attention due to its potential to revolutionize various industries, including tourism. In the context of tourism, the metaverse can offer new and exciting opportunities. It could enable virtual travel experiences, where users can explore destinations, landmarks, and cultural sites from the comfort of their homes. This could be particularly appealing to individuals who are unable to travel physically, whether due to financial limitations, health concerns, or other constraints. Furthermore, the metaverse could enhance the pre-travel planning phase by providing immersive and realistic simulations. Travelers could virtually explore hotels, resorts, and attractions to make more informed decisions about their travel arrangements. This virtual experience could help manage travelers' expectations and improve overall satisfaction with their chosen destinations.

The metaverse might also foster virtual social interactions among travellers. People could connect with fellow travellers, share experiences, and exchange tips and recommendations. This social aspect of the metaverse could enhance the sense of community among travellers, even if they are physically located in different parts of the world. Additionally, the metaverse could create new business opportunities for the tourism industry. Virtual travel agencies and tour operators could emerge, offering virtual travel packages and experiences. Destination marketing

DOI: 10.4324/9781003497004-2

organisations and tourism boards could leverage the metaverse to promote their locations and attract visitors.

It is worth noting that the concept of the metaverse is still in its early stages, and its full potential and impact on tourism are yet to be realized. However, it presents intriguing possibilities for transforming the way we experience and engage with travel and tourism. This literature review explores the emerging concept of the metaverse and its potential implications for the future of the hospitality industry. The metaverse refers to a virtual reality space that allows users to interact with a computer-generated environment and other users in real time. This review aims to analyze existing literature to understand the role of the metaverse in enhancing guest experiences, transforming business operations, and redefining the concept of physical space within the hospitality sector. Additionally, it highlights the challenges and considerations associated with the implementation of metaverse technology in the industry. The findings of this literature review provide valuable insights into the opportunities and potential future scenarios that the metaverse offers for the hospitality sector. The concept of the metaverse refers to a virtual reality space where users can interact with a computer-generated environment and other users in real time. It is often described as a fully immersive and interconnected virtual world that goes beyond the traditional boundaries of the Internet. The metaverse has gained significant attention in recent years due to advancements in virtual reality, augmented reality, and other immersive technologies. About the future of hospitality, the metaverse presents intriguing possibilities for the industry. Some studies and discussions have explored how the metaverse could impact various aspects of hospitality, including customer experiences, marketing, communication, and even the design and operation of physical spaces.

Review of Literature

The metaverse concept has gained significant attention in recent years, promising to revolutionize various industries, including tourism. This literature review aims to explore the role of the metaverse in the future of tourism by examining relevant studies on technology, virtual reality, augmented reality, online reviews, and the tourist experience. The selected articles provide insights into the potential impacts, challenges, and opportunities associated with the metaverse in the context of tourism. In recent years, the tourism industry has witnessed a significant transformation due to advancements in technology. This chapter aims to explore the importance of dynamic tourism experiences and the role of technology in enhancing them. We will delve into the potential of VR and AR in creating immersive and interactive experiences for tourists. Additionally, we will discuss the concept of intelligent tourism and its relationship with the metaverse, as proposed by Buhalis and Sinarta (2019). Finally, we will examine the enablers and barriers of the tourist experience within the context of technological advancements, as highlighted by Neuhofer et al. (2015). To provide a comprehensive understanding, we will also refer to the work of Guttentag (2010) on virtual reality applications and their implications for tourism.

The Importance of Dynamic Tourism Experiences

Oskam and Boswijk (2016) emphasize the significance of dynamic tourism experiences in capturing the attention of modern travellers. Traditional sightseeing and passive engagement are no longer sufficient to meet the evolving expectations of tourists. Dynamic experiences involve active participation, interactivity, and personalization, enabling tourists to immerse themselves in the destination's culture and environment. By leveraging technology, destinations can offer innovative and captivating experiences that go beyond the ordinary.

Enhancing Dynamic Experiences With VR and AR

VR and AR technologies have emerged as powerful tools in shaping dynamic tourism experiences. Oskam and Boswijk (2016) suggest that VR and AR have the potential to create immersive environments where tourists can virtually explore destinations, interact with virtual objects, and even relive historical events. These technologies can transport users to different times and places, allowing them to experience destinations uniquely and engagingly. VR and AR also enable tourists to visualize accommodations, attractions, and activities before making their booking decisions, leading to more informed choices and increased satisfaction.

Intelligent Tourism and the Metaverse

Buhalis and Sinarta (2019) introduce the concept of intelligent tourism, which integrates smart technologies and data analytics to provide personalized and context-aware experiences. They propose that the metaverse, a collective virtual shared space, can serve as a platform for intelligent tourism. In the metaverse, tourists can engage with intelligent avatars, receive personalized recommendations, and access real-time information based on their preferences and location. This integration of technologies and data can enhance the overall tourism experience by tailoring it to individual needs and enabling seamless interactions with the destination.

Enablers and Barriers in Technological Advancements

Neuhofer et al. (2015) shed light on the enablers and barriers associated with technological advancements in the tourism industry. They identify VR and AR as transformative technologies that have the potential to enhance the tourist experience significantly. These technologies offer opportunities for destination marketing, education, and destination management. However, challenges such as high implementation costs, limited user acceptance, and technical constraints need to be addressed. Overcoming these barriers requires collaboration between tourism stakeholders, technological innovators, and policymakers to ensure the successful integration of these technologies into the tourism ecosystem.

Implications and Future Directions

Guttentag (2010) provides insights into the applications and implications of VR in tourism. He discusses various VR applications, including virtual tours, virtual museums, and virtual events. Guttentag highlights that VR can enhance the competitiveness of tourism destinations, create new revenue streams, and offer opportunities for sustainable tourism. As technology continues to advance, future research should focus on addressing the challenges associated with VR and AR adoption, exploring innovative ways to integrate intelligent technologies into the tourism experience, and understanding the long-term effects on sustainability and cultural authenticity. Technology plays a vital role in shaping dynamic tourism experiences. The potential of VR and AR in creating immersive and interactive environments has been recognized by researchers and practitioners alike. The concept of intelligent tourism, coupled with the metaverse, offers new possibilities for personalization and context awareness. However, the successful integration of these technologies requires careful consideration of enablers and barriers. Stakeholders within the tourism industry must collaborate to overcome challenges and unlock the full potential of these transformative technologies. As we look to the future, continued research and innovation are crucial to understanding the implications of technology on tourism and ensuring sustainable and authentic experiences for travelers. The rapid advancements in technology have transformed the way people travel and seek information about potential destinations. Two papers, Ouyang et al. (2017) and Kim and Lee (2018), shed light on different aspects of travelers' behavior and its implications for destination choice and experiential involvement. While Ouyang et al. (2017) explore the antecedents and outcomes of travelers' information-seeking behavior, Kim and Lee (2018) investigate the role of AR in enhancing tourists' experiential involvement. This chapter aims to synthesize the findings of these two studies and explore how their insights can be applied to designing immersive metaverse experiences for travelers.

Information-Seeking Behavior and Destination Choice

Ouyang et al. (2017) delve into the factors that influence travellers' information-seeking behaviour and its impact on destination choice. The study identifies several antecedents, such as travel motivation, perceived risk, and past travel experience, that drive individuals to actively seek information. Additionally, the authors highlight the outcomes of information-seeking behaviour, including improved decision-making, satisfaction, and destination loyalty. These findings underline the importance of providing comprehensive and easily accessible information to travelers in the metaverse.

Augmented Reality and Experiential Involvement

Kim and Lee (2018) focus on the role of AR in smart tourism destination marketing and its influence on tourists' experiential involvement. The study demonstrates that AR can enhance tourists' engagement and immersion by providing interactive and personalized experiences. Through AR applications, travelers can explore destinations virtually, visualize attractions, and access real-time information. AR not only enriches travelers' experiences but also empowers them to actively participate in shaping their own travel narratives.

Integrating Information-Seeking Behavior and Augmented Reality in the Metaverse

Considering the insights from Ouyang et al. (2017) and Kim and Lee (2018), it becomes evident that integrating information-seeking behaviour and augmented reality can contribute to creating immersive metaverse experiences for travellers. In the metaverse, users can navigate through virtual worlds, interact with digital representations of real and fictional places, and access vast amounts of information. By leveraging the principles identified in these studies, developers can design metaverse experiences that meet the information needs of travellers while enhancing their engagement and experiential involvement.

Personalization and Contextualization in the Metaverse

One key aspect that emerges from both papers is the importance of personalization and contextualization in travelers' experiences. Ouyang et al. (2017) highlight the role of individual preferences and motivations in information-seeking behavior, suggesting that tailored content can better meet the needs of diverse travelers. Similarly, Kim and Lee (2018) emphasize the potential of AR in delivering personalized and contextually relevant information to tourists. In the metaverse, developers can leverage user data and artificial intelligence (AI) algorithms to create customized experiences that align with individual preferences, enhancing the overall satisfaction and engagement of travelers. The studies by Ouyang et al. (2017) and Kim and Lee (2018) provide valuable insights into travelers' information-seeking behavior and the role of augmented reality in enhancing tourists' experiences. By integrating these findings, developers can design immersive and personalized metaverse experiences that cater to the diverse needs of travelers. However, it is essential to address challenges related to information accuracy, privacy, and inclusivity to ensure that the metaverse becomes a platform that empowers and enriches travelers' experiences responsibly and ethically. The synthesis of research in this chapter contributes to a deeper understanding of how the metaverse can be leveraged to revolutionize the way people travel, explore destinations, and engage with their surroundings. The emergence of the metaverse, a virtual reality-based digital universe, has transformed various industries, including hospitality and tourism. Understanding the impact of online reviews and virtual reality on user perceptions and decision-making processes in the metaverse is crucial for businesses operating in this new paradigm. This chapter aims to explore the implications of the aforementioned articles for reputation management and user experience in the metaverse.

Online Review Platforms and Social Media Analytics

Xiang et al. (2017) conducted a comparative analysis of major online review platforms, such as TripAdvisor, Yelp, and Google Reviews, and their implications for social media analytics in the hospitality and tourism industry. The study highlighted the importance of monitoring and analyzing user-generated content for businesses. In the metaverse, user feedback and reputation management are equally important, as users heavily rely on online reviews to make informed decisions about virtual experiences.

The Influence of Online Reviews in the Metaverse

Drawing from Xiang et al. (2017), online reviews play a significant role in shaping perceptions and expectations of virtual destinations in the metaverse. Positive reviews can enhance the reputation of virtual experiences and attract more users, while negative reviews can deter potential visitors. Thus, businesses in the metaverse should actively manage their online presence and engage with users to ensure positive user experiences and maintain a favourable reputation.

Virtual Reality and Destination Image Formation

McFee et al. (2019) investigated the effects of VR on destination image formation in the physical tourism context. They found that immersive experiences through VR positively influence tourists' perceptions of destinations, enhancing their desire to visit. This finding has implications for the metaverse, where users' experiences are predominantly virtual. VR technologies in the metaverse can significantly impact users' perceptions of virtual destinations, influencing their decision-making and engagement with virtual experiences.

The Role of Virtual Reality in the Metaverse

In the metaverse, virtual reality technologies provide users with immersive and realistic experiences. Based on the findings of McFee et al. (2019), we can infer that VR in the metaverse has the potential to shape users' perceptions of virtual destinations, influencing their desire to engage with specific experiences. Businesses operating in the metaverse can leverage VR technologies to create compelling virtual environments and attract a larger user base.

Implications for the Metaverse

Both articles provide valuable insights into the metaverse, despite not directly addressing it. Xiang et al. (2017) emphasize the importance of user-generated content and reputation management, which are equally relevant in the metaverse.

McFee et al. (2019) highlight the potential of VR technologies in shaping users' perceptions and decision-making processes, which can be applied to virtual experiences within the metaverse. Combining these insights, businesses in the metaverse should actively monitor online reviews, engage with users, and utilize VR technologies to enhance destination image and attract users. Understanding the influence of online reviews and virtual reality in the hospitality and tourism industry provides valuable insights for managing reputation and enhancing user experiences in the metaverse. By leveraging user-generated content and immersive technologies, businesses can create compelling virtual experiences that shape user perceptions and drive engagement in the metaverse. Further research is required to explore the full potential of these insights in the evolving metaverse landscape (Xiang et al., 2017; McFee et al., 2019).

The Advent of the Metaverse Has Revolutionized the Way Individuals Interact With Virtual Environments, Including Travel Experiences

As the metaverse continues to grow in popularity, understanding the factors that shape tourists' decision-making processes and perceptions within this digital realm becomes crucial. This chapter aims to explore the role of online travel reviews, uncertainty, and conformity, as well as social media in influencing tourists' behaviour and perceptions in the metaverse.

Perceived Usefulness of Online Travel Reviews: The Role of Uncertainty and Conformity

The study by Gursoy et al. (2015) investigates the perceived usefulness of online travel reviews and how uncertainty and conformity affect tourists' reliance on these reviews. The findings suggest that online travel reviews play a significant role in shaping tourists' decision-making processes, with perceived usefulness being a key determinant. Uncertainty and conformity are identified as important factors that influence tourists' trust in online reviews. The study highlights the need for trustworthy and informative metaverse experiences to enhance tourists' reliance on online reviews.

Leveraging Uncertainty and Conformity in the Metaverse

Drawing from Gursoy et al.'s (2004, 2015) findings, it becomes essential to design metaverse experiences that address tourists' uncertainty and conformity needs. Strategies such as providing accurate and detailed information, showcasing social proof, and utilizing influencers can help minimize uncertainty and foster conformity among metaverse tourists. By addressing these factors, destination marketers can enhance the perceived usefulness of online travel reviews and increase tourists' trust in the metaverse.

The Influence of Social Media on Tourists' Perceptions in the Metaverse

The research of Xiang et al. (2015) delves into the impact of social media platforms on travellers' perceptions, focusing on a case study of Australia. The study reveals that social media plays a pivotal role in shaping tourists' destination perceptions, with factors such as visual content, user-generated reviews, and interaction with peers influencing their decision-making processes. Understanding the mechanisms by which social media shapes perceptions can provide valuable insights for destination marketing and engagement in the metaverse.

Leveraging Social Media Mechanisms in the Metaverse

Building upon Wang et al.'s (2014) and Li et al.'s (2020) findings, it becomes crucial to leverage similar social media mechanisms within the metaverse for destination marketing and engagement. Creating visually appealing and immersive experiences, encouraging user-generated content and engagement, and utilizing influencers can help shape tourists' perceptions in the metaverse. By integrating social media strategies effectively, destination marketers can enhance tourists' engagement, foster positive perceptions, and facilitate decision-making processes within the metaverse.

Research Methodology

Bibliometric Search

Early in 2023, we searched for publications to be included in the review using a four-stage search approach that we developed: database search, academic filtration, language filtration, and subject filtration (Figure 1.1).

Stage 1 – Search for Research Papers From the SCOPUS Database: Scopus is a comprehensive scientific database that is widely used for bibliometric evaluations due to its coverage of high-quality and relevant scientific publications. It is ideal for projects that require a large corpus of data for analysis and review. While it may have less coverage than some other databases, its measures are highly correlated with those from alternative databases such as Web of Science. Therefore, it is a valuable resource for researchers and academics seeking to assess the impact and quality of scientific publications. The Scopus database was chosen for its comprehensiveness of bibliometric data for the publications it indexes and coverage of articles that matched a stringent set of indexing requirements (e.g., scientifically and academically relevant). Scopus is a scientific database widely recommended for bibliometric evaluations (Donthu et al., 2021; Paul et al., 2021). Indeed, it is ideal for projects aimed at curating a large corpus for review (Paul et al., 2021). Although the latter has less coverage than the former, the Scopus database has been recognized as a high-quality source of bibliometric data (Baas et al., 2020), and the correlation of its measures with those available from alternative scientific databases such as Web of Science has been demonstrated as "extremely high" (Archambault et al., 2009).

PRISMA 2009 Flow Diagram

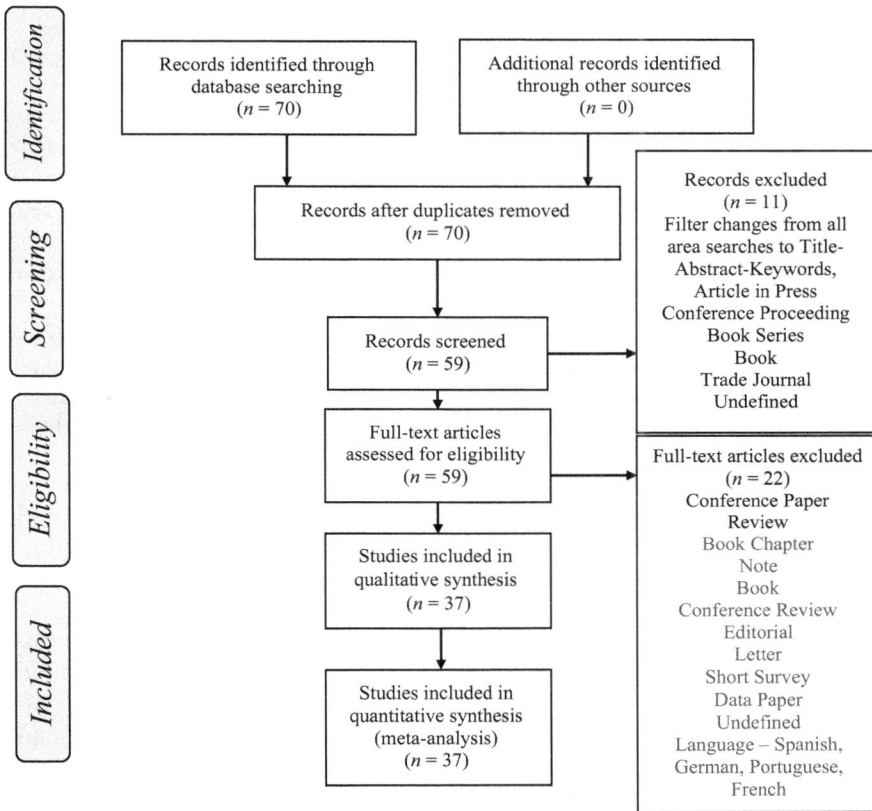

Figure 1.1 PRISMA flow chart for bibliometric review (Moher et al., 2009).

Stage 2 – Scholarly Filtration: Only journal papers and conference proceedings were chosen for inclusion since they are often (1) assessed for originality and (2) submitted for thorough peer review.

Stage 3 – The Application of the Language Filter: We decided to add English-language items that were only kept for Stage 2. Due to the following factors: (1) We are native English speakers.

Stage 4 – The Application of the Subject Filter: We choose to include articles with the following keywords: (1) Metaverse, (2) Tourism, and (3) Hospitality.

In the first search of literature in the SCOPUS database, a total of 70 papers were found. The formula used for searching the research papers in SCOPUS database is as follows: TITLE-ABS-KEY (((("Metaverse") AND ("tourism" OR "HOSPITAL-ITY"))) AND (LIMIT-TO (DOCTYPE, "ar")) AND (LIMIT-TO (LANGUAGE, "English")) AND (LIMIT-TO (SRCTYPE, "j")).

The article search for literature review was conducted in 2023, and the search results were guided by strategic development based on four steps: searching papers in the database and filtration of article results based on language and relevant subjects.

Stage 1 is based on a database search on the Scopus database using the relevant keywords. The Scopus database was selected due to the popular scientific relevance of this database. This database includes comprehensive bibliometric information with high-quality data sources for literature review. In Stage 1, 70 papers were provided by the Scopus database based on our search commands.

Stage 2 is based on removing duplicate research papers. In this study, only the Scopus database was used for article search, so no duplicate articles were available from another database, but 11 articles were removed based on their, multiple availability in the search results. After Stage 2, a total of 59 papers were included for further filtration.

Stage 3 is based on filter applications. At this stage, only qualified papers were included. Multiple filters were applied based on the quality and novelty of the research papers. Some of the filter criteria for exclusion are as follows: Filter changes from all area searches to Title-Abstract-Keywords, Article in Press, Conference Proceeding, Book Series, Book, Trade Journal, and Undefined. At Stage 3, a total of 22 articles were removed based on the above filtration criteria. After Stage 3 filtration, a total of 37 research papers were used for further filtration stages.

Stage 4 is based on the application of subject filtration, and language filtration with another important exclusion filter such as Conference Paper, Review, Book Chapter, Note, Book, Conference Review, Editorial, Letter, Short Survey, Data Paper, Undefined, and Language – Spanish, German, Portuguese, and French. At Stage 4, zero articles were removed using the above filter criteria. At the final stage, a total of 37 articles were used for the bibliometric analysis.

Results

The data analysis results are presented in the following using figures and tables. Figure 1.2 represents the bibliographic network of the most frequent keywords. In Figure 1.2, metaverse tourism and virtual reality are the most frequent keywords, and also these keywords are making strong associations with other keywords.

Table 1.1 shows the list of most contributing countries with the number of documents published, citations in the domain of tourism, and metaverse. The UK is contributing the highest 154 number of citations with seven documents published. The studies of the UK also have the highest number 1,178 for total link strength. The USA is the second most contributing country with 139 citations and 817 numbers

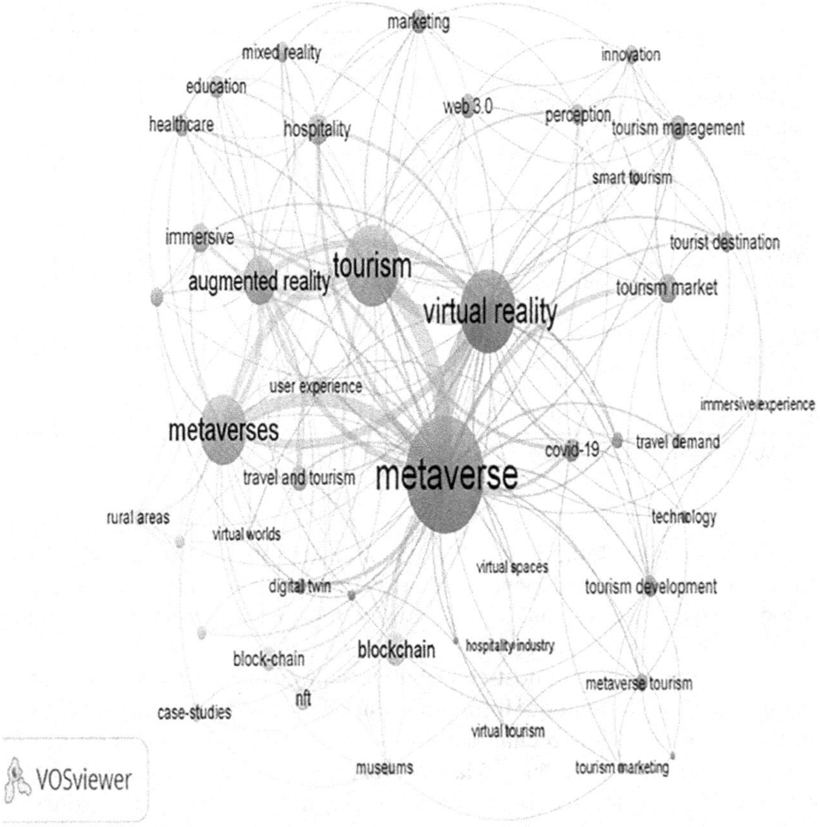

Figure 1.2 Bibliographic network of most frequent keywords analysis of data.

Table 1.1 Most contributing countries

Country	Documents	Citations	Total link strength
UK	7	154	1,178
USA	8	139	817
India	11	129	752
Hong Kong	3	116	567
South Korea	9	76	336
China	11	63	1,120
Taiwan	4	36	479
Malaysia	3	32	783
Spain	3	27	29
Singapore	3	13	81
Macao	5	5	460
Turkey	3	0	88

Source: Voviewer SLR analysis based on the SCOPUS database

Table 1.2 Most contributing journals

Source	Documents	Citations	Total link strength
Journal of Hospitality Marketing and Management	2	95	15
International Journal of Contemporary Hospitality Management	2	75	14
Sustainability (Switzerland)	5	52	5
Current Issues in Tourism	2	32	16
Tourism Review	2	15	13
Springer Proceedings in Business and Economics	5	8	8
Influencer Marketing Applications Within the Metaverse	2	1	20
Information Technology and Tourism	7	0	11
Journal of Hospitality and Tourism Insights	2	0	18

Source: Voviewer SLR analysis based on the SCOPUS database

of total link strength. India is the third most contributing country with 129 citations. Furthermore, Hong Kong, South Korea, China, Taiwan, Malaysia, Spain, Singapore, Spain, Macao, and Turkey are the most contributing countries in the research of tourism and metaverse.

Table 1.2 represents the most contributing journals in the area of tourism and the metaverse. The Journal of Hospitality Marketing and Management is the most contributing journal with 95 citations and two documents. The International Journal of Contemporary Hospitality Management, Sustainability (Switzerland), Current Issues in Tourism, Tourism Review, Springer Proceedings in Business and Economics, Influencer Marketing Applications within the Metaverse, Information Technology and Tourism, and Journal of Hospitality and Tourism Insights are furthermost contributing journals. Table 1.3 represents the most contributing authors in the area of tourism and metaverse.

Table 1.4 represents the most frequent keywords. Metaverse is the most frequent keyword with 39 times occurrences and 121 total link strength. The second most frequent keyword is virtual reality with 23 occurrences. Tourism is the third most frequent keyword with 14-time occurrence. Furthermore, frequent keywords are augmented reality metaverse tourism, blockchain, COVID-19, hospitality, tourism development, tourism market, cultural heritage, digital twin, immersive, smart tourism, tourism management, tourist destination, virtual tourism, blockchain, case studies, education, grounded theory, health care, healthcare, hospitality industry, immersive experience, innovation, marketing, mixed reality, museums, nft, perception, rural areas, second life, sustainability, technology, tourism marketing, travel and tourism, travel demand, user experience, VR, virtual spaces, virtual world, virtual worlds, and web 3.0.

Table 1.3 Most contributing authors

Author	Documents	Citations	Total link strength
gursoy d.; malodia s.; dhir a.	1	95	1
buhalis d.; lin m.s.; leung d.	1	53	13
buhalis d.; leung d.; lin m.	1	41	13
koo c.; kwon j.; chung n.; kim j.	1	32	3
koohang a.; nord j.h.; ooi k.-b.; tan g.w.-h.; al-emran m.; aw e.c.-x.; baabdullah a.m.; buhalis d.; cham t.-h.; dennis c.; dutot v.; dwivedi y.k.; hughes l.; mogaji e.; pandey n.; phau i.; raman r.; sharma a.; sigala m.; ueno a.; wong l.-w.	1	29	1
garrido-iñigo p.; rodríguez-moreno f.	1	27	0
buhalis d.; o'connor p.; leung r.	1	22	15
lee u.-k.	1	19	2
Wei d.	1	17	2
zaman u.; koo i.; abbasi s.; raza s.h.; qureshi m.g.	1	17	2

Source: Voviewer SLR analysis based on the SCOPUS database

Cluster Analysis of Keywords

Cluster 1 – Exploring the Intersection of COVID-19, Digital Twins, and Virtual Realities in Tourism and Sustainability

The following keywords have been segmented in Cluster 1 by the VosViewer software (Table 1.5). Here Cluster 1 represents the Intersection of COVID-19, Digital Twins, and Virtual Realities in Tourism and Sustainability.

The literature review reveals significant implications in the intersection of COVID-19, Digital Twins, and Virtual Realities in the context of tourism and sustainability. First, the COVID-19 pandemic has led to widespread travel restrictions, significantly impacting the tourism industry (Li et al., 2014; Gossling et al., 2020). However, this crisis has accelerated the adoption of digital technologies in tourism, such as Digital Twins and Virtual Realities, which have become essential tools for real-time co-creation and the development of smart tourism in cities (Buhalis & Sinarta, 2019). Virtual reality's role in tourism marketing has been acknowledged, opening up new avenues for engaging tourists (Jin et al., 2019; Valeri, 2024a; Hazy, 2012; Verma & Rao, 2016; Narayanswamy, 2016). Moreover, Digital Twins have gained traction in tourism, promising innovative approaches to destination management (Li et al., 2020). In the context of sustainability, there is a growing need to assess whether tourism in protected areas contributes to conservation (Buckley et al., 2019). As tourism continues to evolve, the integration of these technologies has the potential to shape the future of sustainable tourism and enhance the tourist

Table 1.4 Most frequent keywords

Keyword	Occurrences	Total link strength
Metaverse	39	121
Virtual reality	23	76
Tourism	14	74
Metaverses	12	61
Augmented reality	6	37
Metaverse tourism	6	10
Blockchain	5	24
COVID -19	4	14
Hospitality	4	20
Tourism development	4	13
Tourism market	4	19
Cultural heritage	3	8
Digital twin	3	9
Immersive	3	19
Smart tourism	3	9
Tourism management	3	15
Tourist destination	3	12
Virtual tourism	3	4
Block-chain	2	15
Case studies	2	11
Education	2	14
Grounded theory	2	4
Health care	2	11
Healthcare	2	14
Hospitality industry	2	1
Immersive experience	2	6
Innovation	2	11
Marketing	2	15
Mixed reality	2	12
Museums	2	10
nft	2	15
Perception	2	12
Rural areas	2	9
Second life	2	8
Sustainability	2	9
Technology	2	7
Tourism marketing	2	6
Travel and tourism	2	16
Travel demand	2	8
User experience	2	10
Virtual reality (VR)	2	3
Virtual spaces	2	5
Virtual world	2	6
Virtual worlds	2	4
Web 3.0	2	15

Source: Voviewer SLR analysis based on the SCOPUS database

Table 1.5 Cluster 1 Keywords

Cluster 1 Keywords
COVID-19
Digital twin
Grounded theory
Hospitality industry
Metaverse
Metaverse tourism
Sustainability
Technology
Tourism development
Tourism marketing
Virtual reality
Virtual reality (vr)
Virtual spaces
Virtual tourism
Virtual world
Virtual worlds

Source: Voviewer SLR analysis based on the SCOPUS database

experience while addressing the challenges posed by the ongoing pandemic (Xiang et al., 2015; Wang et al., 2016; Ham et al., 2020).

Cluster 2 (11 Items) – Exploring Augmented Reality, Education, and Healthcare Innovations in Immersive Metaverses for Enhanced User Experiences

The following keywords have been segmented in Cluster 2 by the VosViewer software (Table 1.6). Here Cluster 2 is exploring augmented reality, education, and healthcare innovations in immersive metaverses for enhanced user experiences.

The latest literature review on "Exploring Augmented Reality, Education, and Healthcare Innovations in Immersive Metaverses for Enhanced User Experiences" underscores the profound impact of immersive technologies on education and healthcare. This research has revealed that the integration of AR within immersive metaverses can revolutionize learning experiences by providing interactive, engaging, and personalized educational content (Podsukhina et al., 2022; Valeri, 2024b). Furthermore, it highlights that healthcare applications in metaverses, such as telemedicine and medical training simulations, have the potential to enhance patient care and medical education, ensuring a more accessible and efficient healthcare system. These findings emphasize the transformative potential of immersive metaverses in redefining user experiences, raising the need for continued research and investment in this burgeoning field (Gretzel et al., 2006).

Cluster 3 (Ten Items) – Exploring the Future of Tourism: Immersive Innovation, Smart Marketing, and Web 3.0 Experiences

Table 1.6 Cluster 2 Keywords

Cluster 2 Keywords
Augmented reality
Education
Health care healthcare
Hospitality
Immersive
Metaverses
Mixed reality
Tourism
Travel and tourism
User experience

Source: Voviewer SLR analysis based on the SCOPUS database

Table 1.7 Cluster 3 Keywords

Immersive experience
Innovation
Marketing
Perception
Smart tourism
Tourism management
Tourism market
Tourist destination
Travel demand
Web 3.0

Source: Voviewer SLR analysis based on the SCOPUS database

The following keywords have been segmented in Cluster 3 by the VosViewer software (Table 1.7). Here Cluster 3 is exploring the future of tourism: immersive innovation, smart marketing, and web 3.0 experiences.

The latest literature review on "Exploring the Future of Tourism: Immersive Innovation, Smart Marketing, and Web 3.0 Experiences" reveals significant implications for the tourism industry. With the rapid advancement of immersive technologies such as VR and AR, tourism businesses need to adapt by integrating these innovations into their offerings (Podsukhina et al., 2022). Additionally, the emergence of Web 3.0 and blockchain technologies can enhance trust and transparency in the industry, benefiting both tourists and providers (Mariani & Baggio, 2022). Moreover, the adoption of smart marketing strategies, harnessing data analytics and AI, can enable personalized experiences that cater to individual preferences and needs (Xiang et al., 2021; Santus et al., 2023). As the tourism landscape continues to evolve, embracing these trends will be crucial for staying competitive and meeting the evolving expectations of travelers (Kim & Fesenmaier, 2008).

Cluster 4 (Eight Items) – Exploring the Impact of Blockchain Technology on Cultural Heritage: Non-Fungible Tokens (NFTs), Museums, and Case Studies in Rural Areas and Second Life

Table 1.8 Cluster 4 Keywords

Block-chain
Blockchain
Case studies
Cultural heritage
Museums
nft
Rural areas
Second life

Source: Voviewer SLR analysis based on the SCOPUS database

The following keywords have been segmented in Cluster 4 by the VosViewer software. Here Table 1.8 (Cluster 4 keywords) is exploring the impact of blockchain technology on cultural heritage: NFTs, museums, and case studies in rural areas and second life.

The latest literature review on the impact of blockchain technology on cultural heritage, specifically focusing on NFTs, museums, and case studies in rural areas and Second Life, reveals several significant implications. First, the integration of blockchain and NFTs has the potential to revolutionize the art and cultural heritage sector, offering transparent provenance, authentication, and ownership tracking. This can enhance trust and stimulate increased investment in cultural artifacts. Second, museums can leverage blockchain for digitization efforts, enabling them to create decentralized, immutable records of their collections, thereby preserving cultural heritage for future generations. Furthermore, the case studies in rural areas underscore the role of blockchain in empowering local communities to protect and promote their cultural heritage while also facilitating economic development (Xiang & Gretzel, 2010). Finally, the exploration of Second Life as a virtual cultural heritage space underscores the need for innovative approaches to preserving digital artifacts and experiences, with blockchain offering a promising solution for ensuring the longevity of these virtual heritage sites (Podsukhina et al., 2022; Cao et al., 2023; Shao et al., 2022).

Conclusion, Recommendations, and Future Scope of the Study

This chapter has explored the perceived usefulness of online travel reviews and the role of uncertainty and conformity in tourists' decision-making processes, as well as the influence of social media on travelers' perceptions. By understanding these factors, destination marketers can design trustworthy and informative metaverse experiences and leverage social media mechanisms to shape tourists' perceptions and behaviors. The metaverse presents exciting opportunities for destination marketing, and further research in this area is warranted to uncover the full potential of this digital realm. The concept of the metaverse, a virtual space where individuals interact with digital environments and each other, has gained significant attention in recent years. The paper by Kularatne et al. (2021) explores the use of VR and AR technologies in the hospitality industry. They highlight how these immersive technologies can enhance guest experiences by providing virtual tours of hotel rooms and destinations, enabling customers to make informed decisions (Kim & Fesenmaier, 2008). Moreover, VR and AR can transform marketing strategies by allowing businesses to create interactive and personalized content, leading to increased customer engagement and brand loyalty. Barrera and Shah (2023) delve into the concept of the metaverse and its implications for the future of the hospitality and tourism sectors. They emphasize how the metaverse can revolutionize destination marketing by providing immersive experiences that showcase the unique aspects of various locations. Hotels can leverage the metaverse to enhance their operations, offering virtual check-ins, digital concierge services, and customized experiences for guests. Additionally, the metaverse opens up new opportunities for guest interactions, fostering social connections and community engagement. Sanchez-Amboage et al. (2023) present a comprehensive review of the current research on the intersection of the digital metaverse and the hospitality industry. They highlight various technological advancements and applications, including virtual reality experiences, blockchain implementation, AI-powered customer service, and ethical considerations. Virtual reality experiences can transport guests to distant locations and provide unique and immersive encounters, while blockchain technology ensures secure transactions and transparent operations. AI-powered customer service can offer personalized recommendations and seamless interactions, enhancing guest satisfaction. Ethical considerations related to data privacy, digital divide, and equitable access to the metaverse are also discussed (Kourouthanassis et al., 2017). By examining the findings from the three papers, we can identify several synergies and challenges in the integration of the metaverse and the hospitality industry. Synergies include enhanced guest experiences through immersive technologies, personalized marketing strategies, and increased revenue streams through innovative offerings. Challenges encompass technological infrastructure requirements, data privacy and security concerns, digital literacy and accessibility issues, and the need for industry-wide collaboration and standardization. The metaverse has the potential to transform the hospitality industry by providing immersive experiences, reimagining marketing strategies, and revolutionizing guest interactions. The synthesis of the three papers demonstrates the significant implications of virtual reality, augmented reality, and the digital metaverse on the hospitality sector. As the industry embraces the opportunities presented by the metaverse, stakeholders must navigate the associated challenges to ensure an inclusive and ethical integration. By leveraging the insights from the reviewed papers, industry players can position themselves at the forefront of this digital revolution, shaping the future of hospitality in the metaverse.

In this study, we have examined four distinct clusters of research, each shedding light on various facets of the digital landscape's impact on different industries. Cluster 1 has illuminated the intersection of COVID-19, digital twins, and virtual realities in the context of tourism and sustainability. Cluster 2 has delved into augmented reality, education, and healthcare innovations within immersive metaverses. Cluster 3 has explored the future of tourism in the context of immersive innovation, smart marketing, and Web 3.0 experiences. Cluster 4 has investigated the impact of blockchain technology on cultural heritage, with a focus on NFTs, museums, and case studies in rural areas and Second Life. Through these clusters, we have gained insights into the transformative potential of these digital technologies across diverse domains. For instance, within the hospitality industry, VR and AR have emerged as powerful tools to enhance guest experiences and revamp marketing strategies (Kularatne et al., 2021). The metaverse, as discussed by Barrera and Shah (2023), is poised to revolutionize destination marketing, offering immersive experiences and reshaping the hospitality landscape.

Implications of the Study

The implications of these findings extend far beyond their respective domains. They underscore the imperative for businesses and industries to embrace digital transformation actively. For the hospitality sector, adopting VR, AR, and metaverse technologies can lead to more engaging and personalized guest experiences, ultimately boosting customer satisfaction and loyalty (Kularatne et al., 2021). Similarly, the metaverse's potential in tourism marketing (Barrera & Shah, 2023) highlights the need for businesses to stay at the forefront of digital innovation, investing in technologies that can help them showcase the uniqueness of their offerings. Furthermore, the integration of blockchain technology, as discussed in Cluster 4, offers a new dimension to preserving cultural heritage and ensuring transparency in transactions. NFTs and blockchain can reshape the way cultural artifacts are shared, owned, and preserved, with potential implications for museums and rural areas (Cluster 4).

Future Scope of the Study

The future scope of research in these domains is promising and expansive. Building on the insights from these clusters, several avenues for future research become apparent.

Technological Advancements: Continued exploration of emerging technologies like extended reality (XR), AI, and 5G connectivity can yield novel applications within various sectors. For instance, AI-driven personalization in the metaverse and the use of XR for immersive training in healthcare and education are promising areas of study.

Ethical Considerations: As the digital landscape evolves, ethical concerns around data privacy, digital equity, and the ethical use of immersive technologies require in-depth investigation (Sanchez-Amboage et al., 2023). Future research should explore frameworks and best practices to address these concerns.

Cross-Industry Collaboration: Research should focus on how different industries can collaborate within the metaverse to create seamless, cross-domain experiences. This may involve partnerships between tourism and healthcare, education and sustainability, and more.

Standardization and Regulation: The rapid development of metaverse technologies calls for industry-wide standardization and regulatory frameworks to ensure safe, secure, and ethical usage. Future research can contribute to the formulation of such standards.

The studies in these clusters have provided valuable insights into the transformative potential of digital technologies in various industries. These findings not only have immediate implications for businesses but also set the stage for exciting avenues of future research, paving the way for a digital revolution that promises to reshape how we experience and interact with the world. As we move forward, researchers, industry stakeholders, and policymakers need to collaborate and navigate the evolving digital landscape thoughtfully and responsibly.

References

Archambault, É., Campbell, D., Gingras, Y., & Larivière, V. (2009). Comparing bibliometric statistics obtained from the Web of Science and Scopus. *Journal of the American Society for Information Science and Technology, 60*(7), 1320–1326. https://doi.org/10.1002/asi.21062

Baas, J., Schotten, M., Plume, A., Côté, G., & Karimi, R. (2020). Scopus as a curated, high-quality bibliometric data source for academic research in quantitative science studies. *Quantitative Science Studies, 1*(1), 377–386. https://doi.org/10.1162/qss_a_00019

Barrera, K. G., & Shah, D. (2023). Marketing in the metaverse: Conceptual understanding, framework, and research agenda. *Journal of Business Research, 155*(1), 1–19. https://doi.org/10.1016/j.jbusres.2022.113420

Buckley, R., Brough, P., Hague, L., Chauvenet, A., Fleming, C., Roche, E., & Harris, N. (2019). Economic value of protected areas via visitor mental health. *Nature Communications, 10*(1), 1–10. https://doi.org/10.1038/s41467-019-12631-6

Buhalis, D., & Sinarta, Y. (2019). Real-time co-creation and nowness service: Lessons from tourism and hospitality. *Journal of Travel & Tourism Marketing, 36*(5), 563–582. https://doi.org/10.1080/10548408.2019.1592059

Cao, Z. C., Jones, C., & Temouri, Y. (2023). Tax havens and tourism: The impact of the Panama papers and the crowding out of tourism by financial services. *Journal of Travel Research, 10*(2), 1–10. https://doi.org/10.1177/00472875231179395

Donthu, N., Kumar, S., Mukherjee, D., Pandey, N., & Lim, W. M. (2021). How to conduct a bibliometric analysis: An overview and guidelines. *Journal of Business Research, 133*(1), 285–296. https://doi.org/10.1016/j.jbusres.2021.04.070

Gossling, S., Scott, D., & Hall, C. M. (2020). Pandemics, tourism and global change: A rapid assessment of COVID-19. *Journal of Sustainable Tourism, 29*(1), 1–20. https://doi.org/10.1080/09669582.2020.1758708

Gretzel, U., Fesenmaier, D. R., Formica, S., & O'leary, J. T. (2006). Searching for the future: Challenges faced by destination marketing organizations. *Journal of Travel Research, 45*(2), 116–126. https://doi.org/10.1177/0047287506291598

Gursoy, D., Kim, K., & Uysal, M. (2004). Perceived impacts of online travel information on travel planning: A cognitive elaboration perspective. *Journal of Travel Research, 42*(3), 214–224.

Gursoy, D., Kim, K., & Uysal, M. (2015). Perceived usefulness of online travel reviews: The role of uncertainty and conformity. *Journal of Travel Research, 54*(5), 641–654. http://www.ijbts-journal.com/images/column_1432348023/full%20paper%20paris1.pdf

Guttentag, D. (2010). Virtual reality: Applications and implications for tourism. *Tourism Management, 31*(5), 637–651. https://doi.org/10.1016/j.tourman.2009.07.003

Ham, J., Koo, C., & Chung, N. (2020). Configurational patterns of competitive advantage factors for smart tourism: An equifinality perspective. *Current Issues in Tourism, 23*(9), 1066–1072. https://doi.org/10.1080/13683500.2019.1566303

Hazy, J. K. (2012). Leading large: Emergent learning and adaptation in complex social networks. *International Journal of Complexity in Leadership and Management, 2*(1/2), 52–73. https://doi.org/10.1504/IJCLM.2012.050395

Jin, X. C., Qu, M., & Bao, J. (2019). Impact of crisis events on Chinese outbound tourist flow: A framework for post-events growth. *Tourism Management, 74*(1), 334–344. https://doi.org/10.1016/j.tourman.2019.04.011

Kim, H., & Fesenmaier, D. R. (2008). Persuasive design of destination web sites: An analysis of first impression. *Journal of Travel Research, 47*(1), 3–13. https://doi.org/10.1177/0047287507312405

Kim, S. S., & Lee, C. K. (2018). Augmented reality (AR) for smart tourism destination marketing: Enhancing tourists' experiential involvement. *Journal of Travel Research, 57*(6), 803–817.

Kourouthanassis, P. E., Mikalef, P., Pappas, I. O., & Kostagiolas, P. (2017). Explaining travellers online information satisfaction: A complexity theory approach on information needs, barriers, sources and personal characteristics. *Information & Management, 54*(6), 814–824. https://doi.org/10.1016/j.im.2017.03.004

Kularatne, T., Wilson, C., Lee, B., & Hoang, V. N. (2021). Tourists' before and after experience valuations: A unique choice experiment with policy implications for the nature-based tourism industry. *Economic Analysis and Policy, 69*(1), 529–543. https://doi.org/10.1016/j.eap.2021.01.002

Li, M. W., Teng, H. Y., & Chen, C. Y. (2020). Unlocking the customer engagement-brand loyalty relationship in tourism social media: The roles of brand attachment and customer trust. *Journal of Hospitality and Tourism Management, 44*(1), 184–192. https://doi.org/10.1016/j.jhtm.2020.06.015

Li, X. R., Wang, D., Park, S., & Fesenmaier, D. R. (2014). How UGC motivates tourists to contribute: A uses and gratifications approach. *Journal of Travel Research, 53*(4), 534–549.

Mariani, M., & Baggio, R. (2022). Big data and analytics in hospitality and tourism: A systematic literature review. *International Journal of Contemporary Hospitality Management, 34*(1), 231–278. https://doi.org/10.1108/IJCHM-03-2021-0301

McFee, A., Mayrhofer, T., Baràtovà, A., Neuhofer, B., Rainoldi, M., & Egger, R. (2019). The effects of virtual reality on destination image formation. In J. Pesonen & J. Neidhardt (Eds.), *Information and communication technologies in tourism.* Springer. https://doi.org/10.1007/978-3-030-05940-8_9

Moher, D., Liberati, A., Tetzlaff, J., Altman, D. G., & PRISMA Group*. (2009). Preferred reporting items for systematic reviews and meta-analyses: The PRISMA statement. *Annals of Internal Medicine, 151*(4), 264–269.

Narayanswamy, R. (2016). Leadership is not a destination but a place to come from Gandhi's contribution to evolutionary excellence. *International Journal of Complexity in Leadership and Management, 3*(4), 278–283. https://doi.org/10.1504/IJCLM.2016.087151

Neuhofer, B., Buhalis, D., & Ladkin, A. (2015). Technology as a catalyst of change: Enablers and barriers of the tourist experience and their consequences. *Journal of Destination Marketing & Management, 4*(1), 3–11. https://doi.org/10.1007/978-3-319-14343-9_57.

Oskam, J., & Boswijk, A. (2016). The future of tourism experience: Conceptualising and delivering dynamic tourism experiences. *Journal of Travel Research, 55*(2), 140–150.

Ouyang, Z., Gursoy, D., & Sharma, B. (2017). Role of trust, emotions and event attachment on residents' attitudes toward tourism. *Tourism Management, 63*(1), 426–438. https://doi.org/10.1016/j.tourman.2017.06.026

Paul, J., Lim, W. M., O'Cass, A., Hao, A. W., & Bresciani, S. (2021). Scientific procedures and rationales for systematic literature reviews (SPAR-4-SLR). *International Journal of Consumer Studies, 45*(4), 1–16. https://doi.org/10.1111/ijcs.12695

Podsukhina, E., Smith, M. K., & Pinke-Sziva, I. (2022). A critical evaluation of mobile guided tour apps: Motivators and inhibitors for tour guides and customers. *Tourism and Hospitality Research, 22*(4), 414–424. https://doi.org/10.1177/14673584211055819

Sanchez-Amboage, E., Enrique Membiela-Pollán, M., Martínez-Fernández, V. A., & Molinillo, S. (2023). Tourism marketing in a metaverse context: The new reality of European museums on meta. *Museum Management and Curatorship, 38*(4), 468–489, https://doi.org/10.1080/09647775.2023.2209841

Santus, K., Nafi, S., Mallik, N., & Valeri, M. (2023). Mediating effect of emotional intelligence on the relationship between employee job satisfaction and firm performance of small business. *European Business Review, 35*(5), 624–651. https://doi.org/10.1108/EBR-12-2022-0249

Shao, Z., Zhang, L., Brown, S. A., & Zhao, T. (2022). Understanding users' trust transfer mechanism in a blockchain-enabled platform: A mixed methods study. *Decision Support Systems, 155*(1), 1–12. https://doi.org/10.1016/j.dss.2021.113716

Sinha, M., Shekhar., & Valeri, M. (2024, forthcoming). How does entrepreneurship education promote innovation and creativity? Insights from literature review. *International Journal of Technology Enhanced Learning, 16*(1). https://doi.org/10.1504/IJTEL.2023.10055678

Valeri, M. (2024a). *Knowledge management and knowledge sharing. Business strategies and an emerging theoretical field.* Springer. https://doi.org/10.1007/978-3-031-37868-3

Valeri, M. (2024b). *Innovation strategies and organizational culture in tourism. Concepts and case studies on knowledge sharing.* Routledge Publishing. https://books.google.com/books?hl=en&lr=&id=BavoEAAAQBAJ&oi=fnd&pg=PA1846&dq=Valeri,+M.+(2024b).+Innovation+strategies+and+organizational+culture+in+tourism.+Concepts+and+case+studies+on+knowledge+sharing.+Routledge+Publishing.&ots=YEZhRVGAw9&sig=9KWIJtB9lqBB4UovS68eHIJRlso

Verma, P., & Rao, M. K. (2016). Authentic leadership approach for enhancing innovation capability: A theoretical investigation. *International Journal of Complexity in Leadership and Management, 3*(4), 284–300. https://doi.org/10.1504/IJCLM.2016.087114

Wang, D., Xiang, Z., Fesenmaier, D. R., & Law, R. (2014). The role of social media in shaping travelers' perceptions: A case study of Australia. *Journal of Travel Research, 53*(3), 321–343.

Wang, X., Li, X. R., Zhen, F., & Zhang, J. (2016). How smart is your tourist attraction? Measuring tourist preferences of smart tourism attractions via a FCEM-AHP and IPA approach. *Tourism Management, 54*(1), 309–320. https://doi.org/10.1016/j.tourman.2015.12.003

Xiang, Z., Du, Q., Ma, Y., & Fan, W. (2017). A comparative analysis of major online review platforms: Implications for social media analytics in hospitality and tourism. *Tourism Management, 58*, 51–65. https://doi.org/10.1016/j.tourman.2016.10.001

Xiang, Z., & Gretzel, U. (2010). Role of social media in online travel information search. *Tourism Management, 31*(2), 179–188. https://doi.org/10.1016/j.annals.2021.103154

Xiang, Z., Stienmetz, J., & Fesenmaier, D. R. (2021). Smart tourism design: Launching the annals of tourism research curated collection on designing tourism places. *Annals of Tourism Research, 86*(1), 1–7. https://doi.org/10.1016/j.annals.2021.103154

Xiang, Z., Wang, D., O'Leary, J. T., & Fesenmaier, D. R. (2015). Adapting to the internet: Trends in travelers' use of the web for trip planning. *Journal of Travel Research, 54*(4), 511–527. https://doi.org/10.1016/j.annals.2021.103154

2 Critical Aspects of Studies Relating to Virtual Tourism in the Context of Values Resulting From Sustainable Tourism, Nature Tourism, Ecotourism, and Cultural Tourism

Małgorzata Kurleto

Introduction

The progress in the field of information and communication technologies (ICT) during the first two decades of the 21st century has revolutionized many areas of human activity. On one hand, it has caused significant disruption to the tourism industry, but, on the other hand, it has brought benefits to tourism and its users. Virtual tourism has evolved from an ICT technology trend to a mainstream tourism business reality, creating opportunities as well as challenges for both practitioners and researchers in the area (Papathanassis & Buhalis, 2007, p. 1). Although the digital information age has greatly changed global tourism, many authors continue to suggest that digitization is still in its infancy in terms of adoption and value creation in the travel industry (Ozdemir et al., 2023, p. 1). Gössling showed (Figure 4: Four stages of ICT) that, tourism appears to have developed through four stages of ICT adoption – opportunity, disruption, immersion, and usurpation – which reflect on new opportunities and risks, and the need for more critical evaluations of the implications of the ICT economy (Gössling, 2021, p. 743). This remark undoubtedly applies especially to Metaverse tourism and to virtual tourism in its VT and AR varieties. The tourism industry should be prepared for the mutual effects of the development of the metaverse for tourism because these new opportunities to enrich the tourist experience may, as many authors believe, bring about the disintegration of the modern understanding of tourism (Volchek & Brysch, 2023, p. 1). The metaverse is a concept of a three-dimensional (3D) digital world. It consists of virtual spaces that can be explored using an avatar you have created. While some experts see the technological evolution of ICT as an opportunity to develop new offerings, others remain wary or reluctant about tourism's relationship with the metaverse, which intends to implement increasingly immersive technologies to offer tourists experiences that blur the lines between what is real and what's virtual. Many questions are being asked, including the two most important: (1) How could the metaverse take over tourism – a practice that requires physical travel? and (2) Whether and to what extent tourism and the metaverse can cooperate? The health crisis in recent years caused by the pandemic has enabled many travel companies

DOI: 10.4324/9781003497004-3

to increase and maintain the use of technological tools to offer virtual reality tours. Some authors see the metaverse as a way to avoid travel and a way to shift to sustainable tourism, which, while creating opportunities for virtual tourism, may work against traditional forms of tourism. While tourism in the metaverse is not and will not be able to replace outdoor experiences, some professionals in the tourism industry assume the possibility of using it to promote places that can be explored virtually because they are not easily accessible or ignored by tourists. A closer look at different types of tourism undoubtedly gives rise to considering whether their relationship with the metaverse is appropriate. Then the question arises whether and to what extent tourists can be offered certain types of tourism, including, for example, clients of cultural, natural, ecological, or adventure tourism – virtual tourism: (VT) or (AR), as a kind of substitute for real tourism. Considering how popular it is currently to look for and highlight the positive relationship between tourism and the metaverse, this study, in reverse to this fashionable trend, aims to analyze scientific studies that speak critically on this subject. In the literature review, 50 scientific studies relevant to the topic were analyzed (including 36 according to the thematic scheme presented in the annex), which will allow for answering the most important research questions and to verify the research hypothesis that the metaverse should not be treated as a development panacea encompassing all tourism. However, it is appropriate to indicate the tourism area where VR and AR are useful. Fourteen items of literature on selected types of tourism were also examined to highlight the value of real tourism, especially natural and cultural tourism. The study hypothetically assumed that VT and AR will not be able to replace real tourism, that is, the consumption of a tourist product but may encourage its purchase. The analysis used secondary research, among which the study of Verma et al. (2022), "Past, present, and future of virtual tourism – a literature review," bridges this knowledge gap through a comprehensive review of 1652 articles published from 2000 to 2021. It is also worth considering the problem of whether metaverse tourism is a kind of tourist niche. The above review intends to answer the question of whether virtual tourism offers a potential opportunity to move toward sustainable tourism based on modern technology or whether other values are important for sustainable tourism. Apart from this most important question posed in the literature study, other research questions that this chapter will try to answer are as follows: What are the disadvantages of virtual tours? Can virtual tourism be harmful to its participants and to what extent? Is Virtual Reality Harmful to Mental Health? Is the Metaverse unethical? Is further development of the Metaverse and virtual tourism possible and in what direction? and What are the limitations of virtual reality?

The Concept of Metaverse, Virtual Tourism, and Augmented Reality and Their Impact on Tourism

High technologies currently have the most significant impact on the tourism industry in almost every area. AR/VR technologies offer tourists the possibility of telepresence (Cheng et al., 2023; Bec et al., 2021). Virtual tourism is becoming an integrative application of high technologies and the tourism industry. The term

"Metaverse," coined by Neal Stephenson in the novel Snow Crash (1992), refers to a virtual world that exists parallel to the real world (How the Metaverse Will Change the Travel Industry, 2023, p. 1). The metaverse can be defined as a set of virtual shared spaces indexed in the real world and accessible via 3D interaction, which people can explore through the use of VR goggles and other equipment (Aïdi, 2022, p. 1).

A more recent definition of virtual tourism also includes the live broadcasting or streaming of tourism activities (Lu et al., 2022, p. 2). Many definitions limit the term to virtual reality and/or augmented reality experiences, including live or streaming tourism activities (Ryan et al., 2019). With all the news revolving around the Facebook Metaverse, interest has arisen around past attempts at creating virtual worlds, including Second Life, a vast 3D-generated virtual world and platform filled with user-generated content where people can interact with each other in real time (Villar, 2022, p. 2). "Metaverse" is a term that has been rapidly gaining ground in the media landscape ever since Facebook founder Mark Zuckerberg announced the creation of the Meta Group (Aïdi, 2022, p. 2). In the Metaverse, living people are represented by their avatars, which spend their time just like us: working, playing, meeting friends, and, of course, shopping. To enter the meta-world and function there, special sets in VR (virtual reality) or AR (augmented reality) technology are needed. Currently, separate platforms offer virtual worlds called "Metaverses." However, it is still not one universal platform, although optimists assume that the development of the metaverse is heading in this direction because of blockchain technology and the new generation of the Internet, Web3 (Dwivedi et al., 2022).

The metaverse is supposed to be a virtual reality and a platform that will connect the real world with the digital world. It is supposed to be an extension of the current digital space, an alternative world, a place where each of us will be able to have our own digital life and act as an avatar. The goal of Metaversum is to enable everyone to lead a parallel life in the virtual world. The private and business world will probably be transferred to the three-dimensional space, and other people, places, and brands will be present in it. In Metaversum, you will be able to go to universities, work, meet friends, and go on vacation; we will simply transfer life there, just like in Second Life. Metaverse will be a development of this game but on a gigantic and worldwide scale. If the assumptions prove to be true, humanity will face another digital revolution. In principle, the Metaversum, albeit in a very basic form, may already be considered to exist. Certainly, these are activities that bring us closer to spending as much time as possible on the Internet, but this is not the sorely announced revolution. Mark Zuckerberg and Bill Gates also differ on the date of creation of the metaverse. Mark Zuckerberg announced that it would take five to ten years to create the necessary but basic infrastructure for the metaverse to become mainstream (Aïdi, 2022, p. 2). This time is necessary to prepare the 3D world also in terms of technical requirements. A different view is presented by Bill Gates, who announced that in two to three years most of the virtual meetings that we conduct today with the use of ordinary 2D cameras will be carried out in the Metaverse. The boundaries and definition of the Metaverse have no limit. Nobody has seen Metaversum yet, because technological giants are just starting to create it. There are a lot of challenges and areas that need to be adapted to the implementation of the Metaverse. The critical factors for the further development of Metaversum are undoubtedly the availability of equipment, fast and stable infrastructure, global platform integration, and purchasing currency, metaverse, which is rapidly gaining popularity in the media landscape, has created many tourism-related opportunities that various countries and companies are eager to take advantage of (Sear, 2022).

Metaverse tourism results from the interaction between the device and the user who puts himself in the shoes of a tourist. Although the experience is virtual, the senses are triggered by stimulating certain situations that are desirable but unavailable at the moment. Through immersion, virtual reality goggles or haptic sensors allow us to experience things that were previously intangible and reconnect with our senses. Through an avatar, a metaverse user can impersonate a tourist, create a virtual tour route, and interact with other avatars. Metaverse goes beyond AR and VR and offers a transformative virtual world experience with large-scale economic, social, and cultural interactions. Volchek and Brysch (2023, p. 6) emphasize that metaverse tourism can be treated as another niche of tourism. Gursoy et al. (2022) state that a conceptual framework for creating metaverse experiences, identifying research gaps, and proposing agenda items with the potential to significantly benefit hospitality and tourism, industry players.

While tourism in the metaverse cannot replace the outdoor experience, some tourism professionals could leverage it to promote less accessible or overlooked places. Tourists can explore these locations virtually through a conceptual framework for creating metaverse experiences, addressing research gaps, and proposing agenda items that could significantly benefit hospitality and tourism industry players. VR proves to be a highly effective tool for tourists, enabling the creation of a virtual world.

Differentiating between AR and VR, an AR overlays virtual elements onto a real image and enhances the user's perception. Through the application, the camera recognizes objects or markers, overlaying graphics, such as showing the interior of an object directly in the camera's view. These additional elements augment reality and further detach the tourist from the real world (Greenwald, 2021). Special goggles are required to use AR, projecting an image onto the real background and displaying a computer-generated, different reality (Ibid.).

The term "immersion" is crucial in virtual reality, where tourists are entirely immersed, allowing them to look around and interact with the virtual environment.

VR refers to a computer-created, three-dimensional virtual environment that, through the ability to move and interact, impacts at least one of the user's five senses (Guttentag, 2010, p. 638). Three key elements characterize VR: visual elements (stereoscopic vision and the ability to look around), appropriate image quality (sharpness), and interactivity, understood as control over virtual experiences (achieved through motion sensors or joysticks). According to Guttentag's division (2010), VR is applied in six areas related to tourism: tourism planning and regional management, marketing, education, accessibility, and heritage protection.

The metaverse may offer a feasible form of travel replacement and supplementation, encouraging people to use virtual travel on different occasions. Additionally, the metaverse has the potential to empower those unable to experience destinations physically, such as disadvantaged groups (Dwivedi et al., 2022).

Forecasts for the augmented reality market (AR) and virtual reality market (VR) indicate a highly promising market value (Talwar et al., 2022). Even in the most unexpected predictions, this potential was not anticipated to be compounded by the pandemic that significantly affected tourism in 2020. COVID-19 and virtual tourism have propelled the AR and VR markets to reach and sustain very high returns (Ibid.).

The advent of the metaverse enhances the social connections between consumers and suppliers within the tourism industry ecosystem (Buhalis et al., 2023). One significant area transformation is trip planning, as the metaverse provides tools to stimulate travel inspiration. The visualization of destinations and journeys delivers valuable information to potential travelers, inspiring their travel dreams or ideas (Ibid.). Photos and videos can convey to viewers the intangible experiences of past travelers, offering travel inspiration. However, these opinions may not be representative due to the problem of subjective interpretation (Ibid.). The metaverse can be employed to compare hotels and transportation, implying that metaverse tourism can encourage tourists to make bookings. Event planners can leverage the metaverse to promote first-class ticket sales for VIPs. Virtual tourism can transport participants to other places without leaving home. Millions of tourists have taken virtual flights in a virtual world, explored a virtual city, and visited a virtual museum, all from the comfort of their homes.

A prevailing concept suggests that the metaverse is instrumental in reshaping the travel industry. The notion of interactive virtual worlds could alter the way customers interact with different locations and possibly substitute some instances of physical travel without detriment to the industry. For instance, augmented reality applications can furnish information about real-world surroundings, leading to interactive hotel elements, AR-powered tourist destinations, and the use of beacon technology to deliver pertinent push notifications at opportune moments. Modern technologies play an increasingly pivotal role in tourism, both in the planning stages and during journeys. Among the technologies employed in tourism, it is imperative to underscore the significance of the dynamically evolving VR and AR, which can be regarded as the next step in the technological advancement of the tourism industry. The study conducted by Verma et al. (2022) addresses this knowledge gap through a comprehensive review of 1652 articles published from 2000 to 2021. According to this study, the humanization of the travel experience through virtual and augmented reality has gained popularity, but fragmented literature hinders a holistic view. The study confirms, among other things, that virtual reality is an extremely effective tool for tourist planning, enabling the visualization of planned investments created in the virtual world. VR is also employed for cognitive and educational purposes and can be successfully utilized in cultural tourism, such as in museums (Guttentag, 2010). VR provides users with ample opportunities to explore the world without leaving home, contributing to increased accessibility to places of interest for both tourists and researchers, particularly in hard-to-reach areas like diving on the Great Barrier Reef, climbing Mount Everest, visiting a Hawaiian Volcano, venturing into outer space with NASA, or going on safari in Africa. The Immersive Experience Digital Report 2020 underscores that those who believe in the apocalyptic vision of the world being replaced by machines, computers, or extraordinary software are mistaken (Wein, 2021). According to Koo et al. (2022), the core technologies of metaverse tourism will lead to a new level of immersive experiences. The mentioned authors emphasize that tourists can develop more realistic expectations in the pre-trip stage, but virtual tourism can never replace traditional travel (Wein, 2021, p. 2). While there are numerous opportunities to immerse tourists in the world of VR/AR and harness these leading technologies, it is imperative to consider the applications and costs before embarking on the virtual and augmented world journey. VR has been promoted as a means to promote empathy by allowing individuals to virtually experience what it is like to be in someone else's situation. However, meta-analysis studies revealed that "VR was no more effective at increasing empathy than less technologically advanced empathy interventions such as reading about others and imagining their experiences" (Martingano et al., 2021, p. 1). It is noteworthy that designers have introduced new conventions and genres supporting more expressive interpretations of the world. If these expressive artifacts become more realistically spectacular, as CGI effects have in movies over the last three or four decades, the audience's sophistication will similarly increase to fully appreciate and ultimately take for granted the richly fabricated details of characters, believing in the world (Mura et al., 2017, pp. 145–159). Interactive environments necessitate a clearer partnership than mere voluntary suspension of disbelief; they become real by "actively creating beliefs" through evoking and satisfying specific intentional gestures of commitment (Ibid.). As soon as we cease participating due to confusion, boredom, or unpleasant agitation, the illusion dissipates. The barrier is therefore higher for VR than for movies or books because each interaction introduces the possibility of weakening faith.

VR or AR, like any medium, can be used for both benevolent and malevolent purposes. If a tourist wishes to support VR practices serving humanitarian and ethical purposes, they must be grounded in the very real challenges of the human world and free from false and distracting expectations – both fearful and hopeful – regarding a magical escape from our shared physical reality into an improbable virtual future.

Pros and Cons of Virtual Tourism in Contemporary Scientific Works in the Context of the Impact of Modern ICT Technology on Tourism

Pencarelli (2019, p. 456) articulated the trend in the travel and tourism industry, driven by technological advancements, as a new ecosystem of tourism values. This ecosystem encompasses virtualization, decentralization, real-time data gathering and analysis capability, service orientation, and modularity. The dependence of tourism on modern technology is evident, and as emphasized by Stankov and Gretzel (2020, p. 477), various technological innovations such as Big Data Analytics, artificial intelligence (AI), blockchain, location-based services, and virtual and augmented reality systems have been integrated into tourism practices. Duy et al. (2020, p. 2) highlight the significance of the emergence of 5.0 technology and its impact on sustainable tourism. The Symbiont decentralized network for Web 5.0 plays a crucial role in influencing the tourism ecosystem and sustainable development. In Web 5.0, tourists leverage smart communication devices (SC), such as smartphones, phablets, and humanoid robots, along with augmented and AR and VR, to enhance their experiences in the 3D virtual world. While there are numerous arguments favoring virtual tourism, the literature analyzed, including works by Ryan et al. (2019), Negrão (2020), Wein (2021), and Prasanna (2022), predominantly emphasizes arguments against this form of tourism. Proponents highlight the opportunity VR provides to explore any place on Earth, overcoming financial or health limitations and enabling dream trips in the virtual world (Gardonio, 2017). Additionally, virtual travelers can avoid unpleasant experiences such as flight delays, crowded tourist attractions, or language barriers (Pikirayi, 2017). VR's positive impact on safety is notable, as tourists in the virtual world are shielded from various risks, ranging from natural disasters to terrorism, robberies, or thefts. Moreover, the benefits of virtual reality extend to its positive impact on the natural environment. Virtual tourists visiting protected areas neither contribute to environmental degradation nor disrupt valuable resources, allowing them to witness these areas in pristine condition (Making Tourism More Sustainable, 2005). Virtual tourism is often considered an isolating, individual experience, transporting the tourist to a different place away from their existing environment. Virtual tourism boasts a minimal environmental impact, resulting in fewer CO_2 emissions, reduced waste, lower consumption of flora and fauna, and minimal disruption to natural ecology and wildlife. The use of VR and AR minimizes negative social impacts associated with traditional tourism. As a marketing tool, virtual tourism can stimulate real tourism, offering increased freedom and flexibility in terms of time and location. Utheim (2020) underscores the potential of virtual tourism to promote destinations, enhance customer experiences, and generate local revenue. For audiences engaged with VR and AR, the experience is described as an incredibly immersive sensory journey, allowing users to explore artificial environments.

There appear to be more arguments against virtual tourism based on conducted research than in favor of it. Undoubtedly, the complexity of the real world and the unpredictability of events are challenging to replicate in the virtual realm. Another crucial issue is the absence of genuine interaction with the local community at the visited location. It is worth emphasizing that one of the most important dimensions of tourism is its social aspect. When "traveling" in the virtual world, the "tourist" lacks the opportunity to connect with the local population, making them a mere passive observer. Moreover, even a brief stint in the virtual world can lead to adverse effects, such as dizziness, disorientation, balance disorders, headache, or eye pain (LaMotte, 2017; Wein, 2021). These symptoms, akin to motion sickness, undoubtedly diminish the user's satisfaction with the virtual journey. Virtual reality often fosters an isolating, individual experience-transporting users to a different place, detached from their existing environment. This contrasts with real events, where one of the primary goals is to bring people together and encourage group interaction (Negrão, 2020).

It is important to emphasize the following:

- Regardless of how a tourist perceives it, virtuality is merely a reproduction of reality. The tourist views this reality/world through the "eyes" of the creator, potentially missing crucial information.
- Virtual tourism does not yield the economic benefits that traditional tourism does. Traditional tourism provides a substantial income for the host destination, a key reason these areas encourage tourism.
- Virtual tourism is not universally accessible. Not everyone has the digital resources required for its implementation, and many parts of the world lack sufficient Internet connectivity for a satisfactory experience. This limitation is not exclusive to developing countries, as people in most countries may face poor Internet conditions.

Criticism of metaverse tourism revolves around significant health concerns, summarized in the following points: (LaMotte, 2017; Negrão, 2020; Wein, 2021; Prasanna, 2022; Buhalis et al., 2023; Cheng et al., 2023).

1. Virtual Reality Addiction: Instances of teenagers and adults becoming addicted to virtual reality lead to physiological issues and hinder regular studies.
2. Impact on the Real Human Body: Many people who use virtual reality experience many physical problems. Even a short stay in the virtual world can cause negative effects, such as dizziness, nausea, disorientation, balance disorders, headache, or eye pain (LaMotte, 2017, p. 2).
3. One of the effects of visual perception in virtual environments can lead to myopia which currently is a growing problem around the world (Ibid., s. 3).

These mentioned symptoms can certainly reduce the user's satisfaction with the virtual journey.

When critically examining metaverse tourism, it is essential to address ethical considerations. Virtual environments introduce ethical and legal challenges, as they may facilitate illegal, immoral, or unethical behavior, necessitating appropriate legislation and oversight. The complexity is particularly pronounced in tourism, where users traverse borders, and laws may vary based on location. Ethical concerns in social media marketing often mirror issues in non-digital life, encompassing privacy, social, and economic inequalities, accessibility, identity control, and freedom of creative expression.

Considering these factors, it can be concluded that virtual reality cannot replace real tourism. Moreover, it does not pose a threat to the tourism industry; instead, it has the potential to complement it. The distinct challenges and complexities presented by the virtual realm, particularly in terms of ethics and legality, highlight the unique nature of virtual tourism. While it provides novel experiences, it cannot fully replicate the authenticity, cultural exchange, and tangible interactions that define traditional tourism. Virtual tourism should be seen as a supplementary aspect that coexists with and enhances the traditional tourism experience.

Virtual Tourism and the Context of Values Resulting From Sustainable Tourism, Nature Tourism, Ecotourism, and Cultural Tourism

The examination of sustainable tourism in connection with virtual tourism is crucial because metaverse tourism is frequently considered a form of sustainable tourism. This study aims to showcase the superiority of real tourism over virtual tourism, focusing on the values of genuinely sustainable tourism (Making Tourism More Sustainable – A Guide for Policy Makers, UNEP and UNWTO, 2005). In literature, various activities related to caring for the natural environment and the future of the planet exist under different names and types. These activities, distinct from traditional tourism development, aim to ensure the long-term protection and preservation of natural, cultural, and social resources. Their goal is to contribute positively and equitably to economic development and the well-being of individuals living (European Charter for Sustainable Tourism in Protected Areas, 2015). The need to take a closer look at sustainable tourism in connection with virtual tourism is because metaverse tourism is often considered sustainable tourism. This study is aimed at demonstrating the superiority of real tourism over virtual tourism. The study attempted to show the values of truly sustainable tourism. The term ecotourism is often used, understood as a form of active and in-depth sightseeing of areas of outstanding natural and cultural values, which does not destroy the harmony of natural ecosystems and cultural identity of local communities, and provides funds for the protection of these elements (UNWTO, 2018). The condition for its development is the presence of areas with the highest natural values. Ecotourism protects ecosystems, and the cultural identity of the inhabitants influences the processes of educating society and at the same time brings real benefits to the local population by providing funds for the protection of resources (Page & Dowling, 2002). Ecotourism is a form of sustainable tourism, but not all forms of sustainable tourism are synonymous with ecotourism. In naturally valuable areas, nature tourism is developed, which includes all activities aimed at the protection of natural resources, engaging them in any way (Page & Dowling, 2002). Under the strategy for sustainable tourism expressed in the Resolution of the European Parliament of March 25, 2021 (The Resolution . . . 2021, p. 4), taking into account the progress in the field of "soft" mobility, which is a response to the needs of European consumers, options greener and closer to nature tourism are always the best choice. It is difficult to identify the definition of nature tourism as the broadest form related to the use of the benefits of nature. Nature tourism itself is not a separate type of tourism, but it is part of a different form. Elements of nature tourism can be found in active tourism, sightseeing, and ecotourism (Page & Dowling, 2002).

Often it is the "nature" factor that determines the value and nature of tourism, as well as who and where such tourism attracts. Sustainable tourism is adopting it as a phenomenon in which the activities of tourists do not in any way affect the visited environment and do not leave any losses in it, but at the same time bring benefits – both to themselves as tourists and to residents (Baloch et al., 2023).

Sustainable tourism is a very broad term that includes various forms of tourism. It refers to the principle of sustainable development. Sustainable tourism is tourism-friendly to the natural environment (it is aimed at minimizing the negative impact of tourism development on the environment), local communities, tourists, as well as operators providing tourist services (What is sustainable tourism, 2024). The benefits for each party should be equal. Sustainable tourism upholds the ecological, social, and economic integrity of territories. One of the forms of sustainable tourism is ecotourism (a narrower term than sustainable tourism). It refers more to the protection of the environment and acting under the principles of ecology (The Difference between Ecotourism and Sustainable Tourism, 2023). The term "ecotourism" is commonly used to denote an active and thorough exploration of areas with exceptional natural and cultural values, without compromising the harmony of ecosystems or local cultural identity, while contributing funds to protect these elements (UNWTO, 2018). The development of ecotourism requires the presence of areas with the highest natural values, protecting ecosystems, influencing societal education, and providing tangible benefits to local populations through resource protection funding (Page & Dowling, 2002). Although ecotourism is a form of sustainable tourism, it is important to note that not all sustainable tourism forms are synonymous with ecotourism. Nature tourism, found in areas of natural value, encompasses all activities aimed at the protection of natural resources in various ways (Page & Dowling, 2002). Sustainable tourism, as expressed in the European Parliament's Resolution of March 25, 2021 (The Resolution . . . 2021, p. 4), aligns with the evolving trend of "soft" mobility and emphasizes greener and nature-centric tourism as the preferred choice for European consumers. The definition of nature tourism is challenging to pinpoint, as it represents the broadest form related to the utilization of nature's benefits. Nature tourism is not a standalone type but is integrated into different forms, including active tourism, sightseeing, and ecotourism (Page & Dowling, 2002). The "nature" factor often determines the value and nature of tourism, influencing the audience it attracts. Sustainable tourism, as a phenomenon, ensures that tourists' activities have no adverse impact on the visited environment, leaving no harm but bringing benefits to both the tourists and residents involved (Baloch et al., 2023).

Sustainable tourism is a broad term encompassing various forms, rooted in the principles of sustainable development. Its objective is to minimize the negative impact of tourism on the natural environment, local communities, tourists, and service operators while ensuring equitable benefits for all parties. Sustainable tourism places a priority on the ecological, social, and economic integrity of territories. Ecotourism is a specific form of sustainable tourism that places a stronger emphasis on environmental protection and adherence to ecological principles (The Difference between Ecotourism and Sustainable Tourism, 2023). The Global Ecotourism

Network (GEN) defines ecotourism as "responsible travel to natural areas that conserves the environment, sustains the well-being of the local people, and creates knowledge and understanding through interpretation and education of all involved (visitors, staff, and the visited)" (Definition and key concept – Global Ecotourism Network, 2016, p. 1). According to Teneva (2023), real travel destinations offer benefits such as peace of mind, enhanced creativity, improved communication skills, expanded horizons, increased confidence, real-life education, memorable experiences, and self-understanding. Travelers engaging with locals often learn about their thoughts, habits, traditions, and history. The technological orientation of virtual tourism is seen to minimize physical movement (Verma et al., 2022, p. 3). However, there is a segment of travelers who value personal, face-to-face cultural contact more than virtual experiences. The rise of technology-based tourism challenges both cultural and natural tourism, impacting the classic offerings of historic buildings, museums, and galleries in cultural tourism. The evolution of technology-driven tourism and the growing emphasis on individual tourist experiences pose challenges for both cultural and natural tourism. In the domain of cultural tourism, the conventional package of visiting historic buildings, museums, and galleries typically includes a blend of disseminating knowledge about historical events, promoting culture by introducing the public to masterpieces, and assisting in the formation of cultural identity through the thoughtful selection and interpretation of phenomena (Mikos von Rohrscheidt, 2017, p. 3). Heritage interpretation, facilitated by various technologies, provides ample opportunities for creating attractive narratives, meeting the needs of visitors, and developing cultural tourism. Portales et al. (2018) believe that the moment has come for museums, art galleries, libraries, and other similar places to boldly use available modern technology. Richards emphasizes that effective search and gathering of experiences remain fundamental for the functioning of tourist offices, air transport, and hotels (Richards, 2018, pp. 12–21).

Cultural tourism, nature-based tourism, and ecotourism focus on the experiences and sensations of participants. The emotions induced by virtual tourism can never match those stemming from physical presence. Immersive virtual reality is considered an illusion, incapable of replacing the genuine experiences of real travel.

Discussion and Conclusion

There are differing views on virtual tourism in the literature. Some researchers assert that virtual tourism primarily engages sight and hearing, neglecting the multi-sensory bodily experiences sought by tourists, which include stimulation of smell, taste, and touch (Mura et al., 2017). Conversely, others emphasize that visiting new places is not merely entertainment but also an opportunity to discover new facts, acquire knowledge, and gain a better understanding of the world during travel (Dwivedi et al., 2022). The virtual experience is associated with the sense of presence, defined as the degree to which the virtuality of the experience goes unnoticed as a perceptual illusion, while telepresence is linked to a more intense emotional state (Fredericks, 2021]). However, there is a debate about virtual reality researchers regarding whether the concepts of presence and telepresence are interchangeable (Li et al., 2018). Recent progress in virtual reality technology allows virtual tourism suppliers to offer multisensory experiences with both presence and immersion (Buhalis et al., 2023). Yet, studies on emotions caused by virtual and natural conditions reveal significant differences in the profiles of tourists' experiences in real and virtual environments. Chirico and Gaggioli (2019) hold a different opinion on this matter. Bec et al. (2021) argue that virtual tourism can serve as a tool for protection by reducing tourism pressure on unstable destinations. During the pandemic, virtual tourism served as both a marketing tool to promote destinations and a means of providing entertainment. It also helped reinforce stay-at-home orders by offering engaging activities without physically bringing people to destinations (Lu et al., 2022). Some researchers suggest that virtual tourism is better suited for promoting traditional tourism rather than replacing it. Being in a real destination allows tourists to enjoy more holistic experiences (Fennell, 2021, pp. 767–773). Additionally, concerns exist that even a brief stay in the virtual world can have negative effects (LaMotte, 2017), potentially reducing user satisfaction with virtual travel. Simultaneously, there is a fear that the virtualization of experiences may lead to the alienation of tourists.

The analysis of the literature suggests that traditional tourism will never be replaced by Metaverse tourism, virtual tourism, or augmented tourism. However, for those who cannot, or will never be able to, physically visit certain destinations, virtual and immersive tourism can be an alternative, although it will not replace physical and real tourism (Wein, 2021, p. 3). Virtual tourism provides an opportunity to explore places like Mount Everest or polar expeditions that are otherwise unattainable for ordinary tourists without special health conditions and preparations. It may be a suitable choice for virtual space travel. Nevertheless, a real journey involves sensory experiences, encompassing breathtaking views, unknown tastes and smells, and the sensation of warm sand underfoot touched by the sun. When tourists travel physically, they collect stories of great adventures and mishaps, human experiences unique to them, and those they travel with (Mura et al., 2017). Travel and tourism push individuals out of their homes and comfort zones into foreign lands and cultures, fostering daring encounters and conversations with strangers and contributing to the discovery of the world and self-discovery. Limited by its all-content approach, virtual tourism can never replace the unique advantages and benefits of travel, including mental, physical, emotional, and spiritual dimensions. There is no doubt that metaverse and virtual tourism can be harmful to the physical health of participants, especially when individuals become addicted to virtual reality and disengage from the real world. Travel motivated by a desire to learn about nature or culture remains fundamental in the era of unprecedented global accessibility to various places, contributing significantly to the success of the tourism industry. Equally significant is the lack of real interaction with the local community in virtual tourism, which can diminish user satisfaction with virtual travel. Based on these considerations, it can be concluded that virtual reality cannot substitute for real tourism, and it does not pose a threat to the tourism industry; instead, it can complement it. The literature review highlights serious doubts about the suitability of virtual tourism in certain types of tourism, especially in natural or cultural contexts.

Annex

Table 2.1 The researched literature on the subject grouped according to the adopted research objectives

Studies relating to the impact of new information technologies on tourism	Studies relating to the relationship between the metaverse and tourism	Studies relating to virtual tourism in VR and AR.	Studies containing critical aspects regarding virtual tourism
Stankov. U & Gretzel, U. (2020). Tourism 4.0 technologies and tourist experiences: a human-centered design perspective	Sear. J. M. (2022). The Metaverse and Tourism – How the Tourism Sector will benefit from virtual worlds?	Guttentag. D. A. (2010). Virtual reality: Applications and implications for tourism	Mura, P., Tavakoli, R., Sharif, S. P. (2017). Authentic but not too much: exploring perceptions of authenticity of virtual tourism
Duy, T.M. et al. (2020). Study on the Role of Web 4.0 and 5.0 in the Sustainable Tourism Ecosystem of Ho Chi Minh City, Vietnam	Volchek, K. & Brysch, A. (2023). Metaverse and Tourism: From a New Niche to a Transformation	Talwar, S. et al. (2022). Digitalization and sustainability: virtual reality tourism in a post-pandemic world	Wein, J.A. (2021). Virtual tourism never replace traditional travel
Papathanassis, A. & Buhalis, D. (2007). Exploring the ICT revolution and visioning the future of tourism travel and hospitality industries	Gursoy, D., Malodia, S., Dhir, A. (2022). The metaverse in the hospitality and tourism industry	Ryan, Y., Khoo-Lattimore, C. (2019). New realities: a systematic literature review on virtual reality and augmented reality in tourism research	Martingano, J.A. (2021). Virtual Reality Improves Emotional but Not Cognitive Empathy
Cheng, X., Xue, T., Yang, B., Ma, B. (2023). A digital transformation approach in hospitality and tourism research	Koo, C. et al. (2022). Metaverse tourism: conceptual framework and research propositions.	Lu, J. et al. (2022). The potential of virtual tourism in the recovery of tourism industry during the COVID-19 pandemic	Chirico, A. & Gaggioli, A. (2019). When virtual feels real: Comparing emotional responses and presence in virtual and natural environments
Gössling, S. (2021). Tourism, technology and ICT: a critical review of affordances and concessions	How the Metaverse Will Change the Travel Industry (2023).	Fredericks, L. (2021). The Complete Guide to Virtual Tourism	Pikirayi, T. (2017). Will VR-tourism really replace the traditional physical tourism?

(Continued)

Table 2.1 (Continued)

Studies relating to the impact of new information technologies on tourism	Studies relating to the relationship between the metaverse and tourism	Studies relating to virtual tourism in VR and AR.	Studies containing critical aspects regarding virtual tourism
Pencarelli, T. (2019). The digital revolution in the travel and tourism industry	Dwivedi, et al. (2022). Metaverse beyond the hype: Multidisciplinary perspectives on emerging challenges opportunities,	Greenwald, W. (2021). AR vs. VR: What's the Difference?	Negrão, F. (2020). Virtual and Augmented Reality: Pros and Cons
Fennell, D. A. (2021). Technology and the sustainable tourist in the new age of disruption	Aïdi, N. (2022). Tourism and the metaverse: Toward widespread use of virtual travel?	Utheim, H. (2020). What is virtual tourism and when should you make use of it?	Prasanna (2022). Virtual Reality Advantages And Disadvantages
Li, S.H. (2018). New technology and Internet innovation promote the development of tourism industry in the new era	Villar, N. (2022). What Is Second Life? A Brief History of the Metaverse	Verma, S., Warrier, L., Bolia, B., Mehta, S. (2022). Past, present, and future of virtual tourism-a literature review	LaMotte, S. (2017). The very real health dangers of virtual reality
Ozdemir et al. (2023). A critical reflection on digitalization for the hospitality and tourism industry: . . .	Buhalis, D., Leung, D., Lin, M. (2023). Metaverse as a disruptive technology revolutionizing tourism management and marketing	Bec, A. et. al. (2021). Virtual reality and mixed reality for second chance tourism	Gardonio, S. (2017). Will AR/VR Replace Travel & Tourism?

References

Aïdi, N. (2022). *Tourism and the metaverse: Towards widespread use of virtual travel?* Retrieved February 21, 2023, from https://theconversation.com/tourism-and-the-metaverse-towards-a-widespread-use-of-virtual-travel-188858.

Baloch, Q. B., Shah, S. N., Iqbal, N., Sheeraz, M., Asadullah, M., Mahar, S., & Khan, A. U. (2023). *Impact of tourism development upon environmental sustainability: A suggested framework for sustainable ecotourism.* Retrieved February 10, 2023, from https://link.springer.com/article/10.1007/s11356-022–22496-w.

Bec, A., Moyle, B., Schaffer, V., & Timms, K. (2021). Virtual reality and mixed reality for second chance tourism. *Tourism Management, 83.* Retrieved February 20, 2023, from https://doi.org/10.1016/j.tourman.2020.104256.

Buhalis, D., Leung, D., & Lin, M. (2023). Metaverse as a disruptive technology revolutionizing tourism management and marketing. *Tourism Management, 97,* 104724. Retrieved February 23, 2023, from https://doi.org/10.1016/j.tourman.2023.104724.

Cheng, X., Xue, T., Yang, B., & Ma, B. (2023). A digital transformation approach in hospitality and tourism research. *International Journal of Contemporary Hospitality Management*. Retrieved February 25, 2023, from www.emerald.com/insight/content/doi/10.1108/IJCHM-06-2022 -0679/full/html.

Chirico, A., & Gaggioli, A. (2019). When virtual feels real: Comparing emotional responses and presence in virtual and natural environments. *Cyberpsychology, Behaviour, and Social Networking, 22*(3). Retrieved February 28, 2023, from https://doi.org/10.1089/cyber.2018.0393.

Definition and key concept – Global Ecotourism Network. (2016). Retrieved February 19, 2023, from www.globalecotourismnetwork.org/definition-and-key-concepts/.

Duy, N. T., Mondal, S. R., Nguyen, T. T. V., Dzung, P. T., Minh, X. H., & Das, S. A. (2020). A study on the role of web 4.0 and 5.0 in the sustainable tourism ecosystem of Ho Chi Minh City, Vietnam. *Journals Sustainability, 12*(17). Retrieved March 5, 2023, from https://doi.org/10.3390/su12177140.

Dwivedi, K., Hughes, L., Baabdullah, A. M., Ribeiro-Navarrete, S., Giannakis, M., Al-Debei, M. M., Dennehy, D., Metri, B., Buhalis, D., Cheung, C. M. K., Conboy, K., Doyle, R., Dubey, R., Dutot, V., Felix, R., Goyal, D. P., Gustafsson, A., Hinsch, C., Jebabli, I., . . . Wamba, S. F. (2022). Metaverse beyond the hype: Multidisciplinary perspectives on emerging challenges, opportunities, and agenda for research, practice and policy. *International Journal of Information Management, 66*, 1–55. https://doi.org/10.1016/j.ijinfomgt.2022.102542

European charter for sustainable tourism in protected areas. (2015). The Charter. Retrieved March 6, 2023, from www.europarc.org/wp-content/uploads/2015/05/2010-European-Charter-for-Sustainable-Tourism-in-Protected-Areas.pdf.

Fennell, D. A. (2021). Technology and the sustainable tourist in the new age of disruption. *Journal of Sustainable Tourism, 29*(5), 767–773. Retrieved February 24, 2023, from https://doi.org/10.1080/09669582.2020.1769639.

Fredericks, L. (2021). *The complete guide to virtual tourism*. Retrieved February 21, 2023, from www.cvent.com/en/blog/hospitality/virtual-tourism.

Gardonio, S. (2017). *Will AR/VR replace travel & tourism?* Retrieved February 25, 2023, from https://med ium.com/iotforall/will – ar-vr-replace-travel-tourism-662bf9eeb61b.

Gössling, S. (2021). Tourism, technology and ICT: A critical review of affordances and concessions. *Journal of Sustainable Tourism, 29*(5). Retrieved February 20, 2023, from www. tandfonline.com/doi/full/10.1080/09669582.2021.1873353.

Greenwald, W. (2021). *Augmented Reality (AR) vs. Virtual Reality (VR): What's the difference?* Retrieved March 5, 2023, from https://au.pcmag.com/virtual-reality-1/44886/augmented-reality-ar-vs-virtual-reality-vr-whats-the-difference.

Gursoy, D., Malodia, S., & Dhir, A. (2022). The metaverse in the hospitality and tourism industry: An overview of current trends and future research directions. *Journal of Hospitality Marketing & Management, 31*. Retrieved February 18, 2023, from https://doi.org/10.4324/9781003186342 1-8.

Guttentag, D. A. (2010). Virtual reality: Applications and implications for tourism. *Tourism Management, 31*, s. 637–651.

How the metaverse will change the travel industry. (2023). Retrieved February 10, 2023, from www.revfine.com/metaverse-travel/.

Koo, C., Kwon, J., Chung, N., & Kim, J. (2022). Metaverse tourism: Conceptual framework and research propositions. *Current Issues in Tourism*, 1–7. Retrieved March 8, 2023, from www.tandfonline.com/doi/full/10.1080/13683500.2022.2122781.

LaMotte, S. (2017). *The very real health dangers of virtual reality*. Retrieved February 11, 2023, from https://edition. cnn.com/2017/12/13/health/virtual-reality-vr-dangers-safety/index.html.

Li, S. H. (2018). New technology and Internet innovation promote the development of tourism industry in the new era. *Tourism Tribune, 33*(02), 8–11. Retrieved March 6, 2023, from https://doi.org/10.1080/13683500.2021.1959526.

Lu, J., Xiao, X., Xu, Z., Wang, C., Zhang, M., & Zhou, Y. (2022). The potential of virtual tourism in the recovery of tourism industry during the COVID-19 pandemic. *Current Issues in Tourist, 25*. Retrieved March 3, 2023, from https://doi.org/10.1080/13683500. 2021.1959526.

Making tourism more sustainable – A guide for policy makers, UNEP and UNWTO. (2005). Retrieved March 5, 2023, from https://wedocs.unep.org/bitstream/handle/ 20.500.11822/8741/-Making%20 Tourism%20More%20Sustainable_%20A%20Guide% 20for%20Policy%20Makers-2005445.pdf?sequence=3&isAllowed=y.

Martingano, J. A., Hererra, F., & Konrath, S. (2021). Virtual reality improves emotional but not cognitive empathy: A meta-analysis. *Technology, Mind, and Behavior, 2*(1). Retrieved March 4, 2023, from https://doi.org/10.1037/tmb0000034.

Mikos von Rohrscheidt, A. (2017). Zarys problematyki zarządzania w turystyce kulturowej. *Turystyka kulturowa* nr 1 (An outline of management issues in cultural tourism. *Cultural Tourism*, (1), 8–55.

Mura, P., Tavakoli, R., & Sharif, S. P. (2017). Authentic but not too much: Exploring percep-tions of authenticity of virtual tourism. *Information Technology & Tourism, 17*(2), 145–159. Retrieved March 1, 2023, from https://doi.org/10.1007/s40558-016-0059-y.

Negrão, F. (2020). *Virtual and augmented reality: Pros and cons.* Retrieved March 7, 2023, from www.encora.com/insights/virtual-and-augmented-reality-pros-and-cons.

Ozdemir, O., Dogru, T., Kizildag, M., & Erkmen, E. (2023). A critical reflection on digi-talization for the hospitality and tourism industry: Value implications for stakeholders. *International Journal of Contemporary Hospitality Management, 35*(9), 3305–3321. Retrieved February 11, 2023, from https://doi.org/10.1108/ijchm-04-2022-0535.

Page, S. J., & Dowling, R. K. (2002). *Ecotourism*. Pearson Education Ltd.

Papathanassis, A., & Buhalis, D. (2007). Exploring the information and communication technologies revolution and visioning the future of tourism, travel and hospitality indus-tries, 6th e-tourism futures forum: ICT revolutionising tourism 26–27 March 2007, Guild-ford. *International Journal of Tourism Research, 9*(5), 385–387.

Pencarelli, T. (2019). The digital revolution in the travel and tourism industry. *Informa-tion Technology & Tourism, 22*(5), 455–476. Retrieved March 10, 2023, from https://doi. org/10.1007/s40558-019-00160-3.

Pikirayi, T. (2017). *Will VR-tourism really replace the traditional physical tourism?* Retrieved February 17, 2023, from www.techzim.co.zw/2017/08/will-vr-tourism-really-replace-the-actual-physical-tourism/.

Portales, C., Rodrigues, J. M. F., Goncalves, A., Alba, E., & Sebastian, J. (2018). Digital cultural heritage. *Multimodal Technologies and Interactions, 2*(3), 58. Retrieved March 7, 2023, from https://doi.org/10.3390/mti2030058.

Prasanna. (2022). *Virtual reality advantages and disadvantages what is virtual reality (VR)? Benefits, drawbacks, pros, and cons.* Retrieved February 18, 2023, from www.aplustop per.com/virtual-reality-advantages-and- disadvantages/.

Richards, G. (2018). Cultural tourism: A review of recent research and trends. *Journal of Hospitality and Tourism Management, 36*, 12–21. Retrieved February 18, 2023, from https://www.sciencedirect.com/science/article/abs/pii/S1447677018300755.

Ryan, Y., & Khoo-Lattimore, C. (2019). *New realities: A systematic literature review on virtual reality and augmented reality in tourism research.* Retrieved February 14, 2023, from https://research-repository.griffith. edu.au/bitstream/handle/10072/371111/Yung-PUB4708.pdf.

Sear, J. M. (2022). *The metaverse and tourism – how the tourism sector will benefit from virtual worlds?* Retrieved February 2, 2023, from www.linkedin.com/pulse/ metaverse-tourism-how-sector-benefit-from-virtual-worlds-sear.

Stankov, U., & Gretzel, U. (2020). Tourism 4.0 technologies and tourist experiences: A human-centered design perspective. *Information Technology & Tourism, 22*, 477–488.

Talwar, S., Kaur, P., Nunkoo, R., & Dhir, A. (2022, February). Digitalization and sustain-ability: Virtual reality tourism in a post-pandemic world. *Journal of Sustainable Tourism,*

31(11), 2564–2591. Retrieved February 19, 2023, from https://doi.org/10.1080/0966958 2.2022.2029870.

Teneva, M. (2023). *Top 10 benefits of travelling.* Retrieved February 26, 2023, from https:// skyrefund.com/en/blog/benefits-of-travelling.

The Difference between Ecotourism and Sustainable Tourism. (2023). *GSTC2023 sustainable tourism conference in Antalya, Türkiye.* Retrieved March 3, 2023, from www.gst council.org/gstc2023/.

The Resolution of the European Parliament of 25 March 2021 on the development of an EU strategy for sustainable tourism (2020/2038(INI) (2021/C 494/09). (2021). Retrieved February 21, 2023, from https://eur-lex.europa.eu/legal-content/EN/TXT/PDF/?uri=CEL EX:52021IP0109&from=EN.

UNWTO. (2018). *International seminar on harnessing cultural tourism through innovation and technology.* Retrieved March 10, 2023, from https://webunwto.s3-eu-west-1.amazonaws. com/imported-images/50611/concept_note_final.pdf.

Utheim, H. (2020). *What is virtual tourism and when should you make use of it?* Retrieved February 15, 2023, from https://travelopment.com/what-is-virtual-tourism-and-when-should-you-make-use-of-it/.

Verma, S., Warrier, L., Bolia, B., & Mehta, S. (2022, November). Past, present, and future of virtual tourism-a literature review. *International Journal of Information Management Data Insights*, *2*(2). Retrieved February 8, 2023, from www.sciencedirect.com/science/ article/pii/S2667096822000283.

Villar, N. (2022). *What is second life? A brief history of the metaverse.* Retrieved February 14, 2023, from www.makeuseof.com/what-is-second-life-history-metaverse/.

Volchek, K., & Brysch, A. (2023). *Metaverse and tourism: From a new niche to a transformation.* Retrieved February 24, 2023, from https://link.springer.com/chapter/10.1007/ 978-3-031-25752-0_32.

Wein, J. A. (2021). *Virtual tourism never replace traditional travel.* Retrieved February 9, 2023, from www.tourism-review.com/virtual-tourism-has-grown-in-popularity-news12020.

What is sustainable tourism. (2024). Retrieved March 8, 2023, from www.gstcouncil.org/ what-is-sustainable-tourism/

3 The Future for Business Competitiveness in the Tourism Sector

Tourism Agenda 2030 Perspective

Antonio Emmanuel Pérez Brito*, Martha Isabel Bojórquez Zapata, Adriana Otálora Buítrago, and Marco Valeri

*Corresponding author

Introduction

According to Valeri (2015), advancements in both the economy and society could create novel avenues for commercial ventures. Thus, tourist firms are currently reassessing their conventional technological and organizational frameworks. The correlation between tourist destinations and the environment exemplifies how pertinent sustainability is to the competitiveness of those locations. The interconnectedness between sustainability and competitiveness is predicated on the notion that enterprises that actively pursue improvement in environmental, economic, and social performance gain advantages. In light of the escalating global rivalry, tourist destinations are compelled to adopt a more business-oriented approach than before. The tourism sector has frequently encountered uncertainties and inconsistencies, particularly in the past two decades, and its sustainability and degree of societal dedication have been topics of discussion in various academic fields (Valeri, 2015).

Weaver (2011) posits that the competitiveness of a tourist destination is contingent upon the degree of sustainability exhibited in its tourism development. Sustainability can be determined based on specific, clearly defined factors related to tourist intensity (such as the ratio of tourists to residents, the number of employees, the accommodation capacity, and the average duration of stays) and regulatory measures (including tourism legislation, laws pertaining to the preservation of the natural environment, consideration for the local residents' needs, and the protection of the local cultural heritage).

According to Elmo et al. (2020), sustainable tourism can be conceptualized as a means of resource management that aims to address social, economic, and aesthetic requirements while upholding cultural integrity, biological diversity, and crucial ecological processes. The concept of sustainable tourism exhibits variability in its definition, owing to the diverse characteristics encompassed by the terms "tourism" and "sustainability," and is further shaped by the environmental and political perspectives employed.

DOI: 10.4324/9781003497004-4

Elmo et al. (2020) argue that tourism companies can contribute to sustainability by implementing sustainable processes and practices, as well as by developing and promoting innovative technologies, sometimes referred to as green technologies (e.g., electric vehicles), to address sustainability challenges. Hence, sustainable development can be promoted through the implementation of business processes, the production of corporate products, or a combination of both. The same authors recommend a shift toward more sustainable paths through the adoption of innovative approaches that enhance the capacity to acquire knowledge, effectively govern, and adapt to external stimuli within complex socio-ecological systems. Various theoretical and practical perspectives on sustainability concur that enhancing sustainability requires change, innovation, or adaptation in response to the surrounding environment (Valeri, 2023, 2024). In the contemporary landscape of the tourism industry, the capacity to introduce innovation has become an imperative for enterprises operating within the sector. Such innovation can have two distinct forms: incremental innovation and radical innovation. A firm can integrate sustainability through business model innovation. An innovative and sustainable business model must align the company's profitability with both the economic and non-economic benefits it brings to society (Elmo et al., 2020).

Tourism is a sector within the service industry that relies heavily on natural resources, hence impacting communities residing in tourist locations (Alonso et al., 2023). The idea of overtourism, as discussed by Lew (2020), encompasses a range of negative impacts associated with tourist activities. The hospitality and tourism business has various issues that require attention, particularly in terms of effectively managing its expansion in a sustainable manner to preserve its potential. This necessitates a shift toward more sustainable practices (World Tourism Organization [UNWTO], 2017). According to Falatoonitoosi et al. (2022), sustainable tourism development improves the overall prosperity of the area being targeted. Furthermore, sustainability serves as an indicator of the fundamental characteristics of prosperity, particularly in relation to sociocultural empowerment and environmental quality.

The current discourse revolves around the impact of sustainability on the competitiveness of the tourism industry. Recent research has acknowledged the significance of environmental considerations as a determinant in the decision-making process of individuals when selecting travel options (Alonso et al., 2023). According to Fakfare and Wattanacharoensil (2021), sustainability generates competitive advantages in developing countries. Similarly, Chong and Balasingam (2018) highlight that sustainability enhances tourism growth, especially in sectors such as heritage tourism. Pulido-Fernandez et al. (2019) state that incorporating the concept supports the economic development of tourist sites.

Sustainability is considered important in reducing unfavorable environmental outcomes related to tourism activities. These outcomes include high carbon dioxide emissions, specifically in transportation, tree felling, resource insufficiency, the indiscriminate exploitation of nature, and the production of excessive waste (Gross & Grimm, 2018). To achieve the desired results, the tourism sector must adopt sustainable activities. This leads to developing new consumption habits, adopting new manufacturing techniques, and increasing compliance with environmental regulations. According to Tasci (2017) and Tasci et al. (2021), tourists have knowledge and an understanding of issues such as climate change helps achieve business goals in the field of sustainability.

Tourist destinations offer established tourism products, which are organized by the tourist resources or attractions present in the location (Alves & Ramos, 2015). Hence, while acknowledging that the tension between expansion, particularly in the economic realm, and sustainability, including the contentious notion of sustainable development, transcends the domain of tourism, it is imperative to promptly address this issue within the tourism sector. Inadequately planned tourism development can have profound effects on the resources and attractions of a destination, particularly natural resources and aspects associated with its cultural heritage (Fernandez & Rivero, 2009). Such impacts can cause lasting damage to the destination's reputation, thereby endangering its prospects as a tourist destination or even as a residential area. However, the World Tourism Organization (WTO) acknowledges that the advancement toward sustainable tourism has been sluggish and unsatisfactory. Nonetheless, it remains unclear whether these systems are being properly created, as indicated by Fernandez and Rivero (2009).

The tourism industry assumes a pivotal position in the implementation of Sustainable Development Goals (SDGs) and the attainment of sustainable outcomes. The SDGs, as emphasized by Nunkoo et al. (2021), should not be considered in isolation, but rather as an interconnected set of goals within a comprehensive framework.

From a sociological standpoint, economic activities related to tourism are key to promoting inclusion and solving social concerns at the international level. The various approaches to tourism, for example, social, peace, and solidarity tourisms, play a major role in reducing social differences in societies. Furthermore, tourism is a significant driver of economic growth in various nations, creating considerable employment possibilities and contributing significantly to tax revenues. This phenomenon is especially prevalent in geographically isolated and economically marginalized areas, where the limitation of opportunities for industry diversification hinders the development of a sustainable competitive edge. Tourism also yields other important benefits, for example, it drives infrastructural development, leading to an increased standard of living for the locals. Responsible tourists exhibit a preference for ecotourism sites based on an environmental standpoint, as these sites serve to save the natural surroundings, promote conservation efforts, minimize waste generation, and mitigate pollution. Collaborative consumption, often known as the sharing economy, is experiencing significant growth within the tourism sector, concurrently fostering environmental conservation through efficient resource utilization (Buhalis et al., 2023).

Despite the considerable increase in the publication of articles on sustainable competitive tourism, the link between sustainability, competitiveness, and tourism remains understudied. Furthermore, the industry faces the predicament of meeting the 2030 Agenda, as it has thus far proven unsuccessful in halting environmental deterioration and mitigating social inequalities. This study aims to ascertain the novel strategies that improve/maintain competitiveness in the context of sustainability and the SDGs within tourism enterprises.

This chapter is structured into five discrete sections: Introduction, Literature Review, Study Process, Discussion, and Conclusion.

Literature Review

Competitiveness in the Tourism Sector

According to Salinas et al. (2022), the COVID-19 pandemic has had a particularly detrimental impact on business operations within the tourism industry. Countries whose economies rely heavily on tourism will face a challenging scenario in the years ahead. The restoration of a state of normalcy will be contingent upon the level of competitiveness within the tourism industry.

The level of competition within the tourism industry in any country significantly influences the effectiveness of the sector, its ability to attract a significant number of tourists, and, ultimately, its capability to generate economic prosperity (Guaita et al., 2020). Therefore, in the current context, the level of competitiveness between nations will have a great role so that companies in the tourism industry can recover and re-emerge. The COVID-19 global health crisis exacerbated the disparities between nations, leading to the expectation that countries with higher levels of competitiveness would be better equipped to navigate the consequences of the pandemic (Sigala, 2020).

According to Abreu-Novais et al. (2018), some of the main factors that influence competitiveness in tourism companies are as follows: (1) the charm of the place and the satisfaction of customers, (2) the economic capacity, and (3) the impact of competitiveness associated with the economic development of the local society and the sustainability of businesses.

Due to the current situation that tourism is experiencing, after the challenges it was exposed to during the pandemic are overcome, it is essential to identify the factors that influence the competitiveness of the sector and simultaneously work toward the 2030 Agenda (De Castro et al., 2020).

Ritchie and Crouch (2010) presented a summary of the various factors that affect competitiveness as well as the achievement of the 2030 Agenda in hotel companies, including (1) the perceived quality of the reception at the tourist destination, (2) the promotion and advertising of destination, (3) the safety of the destination, (4) the existing legal regulations and policies, (5) the provision of a cost–benefit analysis to tourists, (6) the image that is offered of the tourist destination, (7) ease of communication, and so on.

According to Mira et al. (2016), the primary factors influencing the competitiveness of tourism enterprises are resource management, destination management models, and cyclical conditions. These factors are also integral components of the 2030 Agenda. Assessing these factors is crucial in generating data to delineate and substantiate tourism policies, territorial planning, the effective engagement of all stakeholders in forging strategic partnerships, the formulation and execution of marketing strategies, the enhancement of product and service quality, and the facilitation of investment attraction.

According to Salinas et al. (2023), three primary determinants influence the competitiveness of tourist enterprises, as well as the realization of the 2030 objective, signifying a forthcoming trajectory for this industry. The primary aspect is the administration and safeguarding of tourist assets, encompassing cultural as well as natural resources. The second aspect pertains to enhancing the transportation and telecommunications infrastructure, and the third aspect pertains to improving the nation's overseas engagement.

Zadeh and Kiliç (2021) affirm that infrastructure contributes to the profitability of the tourism sector; this is why its relevance to the sector is globally recognized. The correlation between the advancement of transportation infrastructure and technology and the notable growth of the tourism industry is evident, as the former leads to an upsurge in tourist arrivals and an overall increase in the tourism gross domestic product throughout various global locations. The competitive performance of the tourism industry is influenced by various factors, including political conditions, the enabling environment, and natural and cultural resources, as identified by these authors. The enhancement of tourism competitiveness and the alignment with the 2030 target primarily hinge upon the implementation of enabling policies and conditions. These entail several factors such as prioritizing travel and tourism, fostering international openness, ensuring pricing competitiveness, and promoting environmental sustainability. An enabling environment, which encompasses factors such as an adequate business climate, health and hygiene standards, safety measures, and the use of information technology and communications, increases the competitiveness of tourism companies. The significance of natural and cultural resources in relation to the performance of tourism cannot be understated.

Ecological Economics

The theoretical framework of ecological economics is grounded in the principle of sustainability and focuses on examining the interconnectedness between the economic system, ecosystems, and the social system. It recognizes the reciprocal influences among these systems and views the economy as an open system. In contrast to orthodox economics, ecological economics acknowledges the finite nature of natural resources and the fragility of the environment. Consequently, many scholars consider it the scientific foundation for achieving sustainable economic practices (Correa, 2006). It acknowledges that the economic system encompasses more than just monetary and exchange value, emphasizing the notion of wealth extending beyond mere capital. In fact, it is grounded in ethical considerations and the utilization of natural resources for their inherent value. The primary aim of this economic system is to promote the collective well-being of society. Furthermore, it is characterized as transdisciplinary, as noted by García (2003). The theoretical and methodological frameworks employed in ecological economics include not only the conventional economic principles of prices, costs, and monetary gains but also analytical elements derived from disciplines such as ecology and thermodynamics.

This integration of diverse perspectives offers a comprehensive understanding of the economic process, rendering it ecologically integrated and substantiating its practical applicability (Castro et al., 2020).

The ecological economy encompasses more than just natural systems; it also tends to human society and contributes to overall well-being, extending beyond economic considerations (Yang et al., 2021). It is crucial to comprehend and actively observe our influence on the natural environment while also ensuring that our activities remain within acceptable limits for the sake of the terrestrial ecosystem. This approach is necessary to foster a harmonious coexistence with the ecosystem, enabling the sustained provision of benefits (Leach et al., 2013). The study of ecological economics involves three fundamental concepts: multidimensional well-being, fair distribution, and sustainable scale (Costanza et al., 2020). The concept of multidimensional well-being highlights that the impact of economic transformations extends beyond mere financial prosperity, covering aspects such as physical health, personal security, and overall life contentment (Chaigneau et al., 2022). The concept of fair distribution refers to the equitable allocation of material output and ecological services among individuals, while the concept of the sustainable scale is employed to contextualize the examination of conventional economic equilibrium within the financial feasibility of ecosystems. Ecological economics is a relatively nascent field (Morgan, 2017), although its fundamental concept aligns with the evolving nature of society and offers several avenues for examining a sustainable future.

From an ecological economics perspective, the advancement of sustainable tourism necessitates the comprehensive consideration of not only the economic and social benefits but also the environmental repercussions associated with tourism. The objective of ecological efficiency in tourism is to attain a sustainable equilibrium among ecology, society, and economy by examining the growth of the tourism sector through the lens of ecological economics (Cheng et al., 2023).

The primary aim of incorporating ecological principles into the tourism industry is to reduce the ecological footprint associated with tourism activities while maximizing the economic gains from tourism and avoiding any adverse effects on the natural environment. This approach seeks to establish a harmonious coexistence between the tourism economy and the ecological environment. Evaluating competitiveness in the context of an ecological economy in tourism involves various variables, including the effective usage of tourism resources, the quality of the ecological environment, and the economic benefits derived from tourism activities. By assessing the ecological economy within the context of tourism, scholars can create valuable sources and directions for the promotion of sustainable growth within the tourism sector (Cheng et al., 2023).

Sustainability and Competitiveness in the Tourism Sector

The global tourism industry is presently encountering significant environmental and competitive obstacles, necessitating complex adaptation decisions from tourism enterprises, destinations, and public officials in the sector. One issue of concern pertains to the substantial expansion of the tourism sector, which has evolved into a formidable industry operating under a growth-oriented paradigm, thus exerting significant pressure on the natural environment. Moreover, the existing design of the system is not viable in light of the challenges posed by climate change and environmental factors. The significant rise in the carbon footprint resulting from tourism activities is worth emphasizing. Specifically, the carbon footprint increased from 5% of the global aggregate in 2005 to 8% in 2013. If this trend persists, a further surge of 40% may be expected by 2025. The adoption of a linear growth model in mass tourism has resulted in the excessive utilization of water and energy resources, as well as the generation of substantial amounts of waste and polluting emissions. Moreover, this approach has contributed to the unsustainable exploitation of the tourism industry, harming competitiveness, biodiversity, cultural heritage, and the overall well-being of local communities (Camisón, 2020). Camisón highlights that the tourism sector is currently at a critical juncture due to increasing public awareness and consumer consciousness regarding environmental issues. In this context, companies and destinations have two options. They can choose to prioritize short-term gains and maintain competitiveness by evading their environmental responsibilities and implementing policies that may degrade natural resources. Alternatively, they can adopt a more responsible approach incorporating sustainable strategies that prioritize the long-term preservation of ecosystems, even if it entails higher costs for their services.

According to Streimikiene et al. (2021), competitiveness and environmental and social concerns can be effectively integrated into sustainable tourism development by implementing innovative strategies and promoting sustainable consumption principles. Additionally, catering to the needs of elderly and disabled individuals through the provision of tailored tourism services can improve the overall well-being of local communities. By aligning with the sustainable development priorities of tourist destinations, these efforts can maximize the positive impacts and benefits derived from sustainable tourism practices. It is imperative to consider the demographic characteristics of tourists and their evolving demands for tourism services and products. Hence, the incorporation of sustainable consumption and social tourism functions must be prioritized in the development of tourism products and services to effectively tackle environmental and social sustainability concerns. The COVID-19 pandemic has exerted a substantial impact on the competitiveness of the tourism sector, presenting novel problems for the advancement of sustainable tourism. Tourism enterprises that have successfully navigated the pandemic will be required to enhance the resilience of their offerings in anticipation of future pandemics, as cautioned by health experts. Additionally, they must develop the ability to adjust to the anticipated change in consumer preferences, characterized by an increased inclination toward sustainable products. In the aftermath of the COVID-19 pandemic, the travel and tourism industry is undergoing significant transformations. These changes are influenced by a variety of factors, including consumer preferences, the availability of destinations, and alterations in regulatory frameworks. Consequently, these issues must be addressed in the future, particularly by promoting sustainable consumption practices so that the sector can enhance its competitiveness and ensure long-term sustainability (Streimikiene et al., 2021).

Methodology

Bibliometrics offers a novel methodology for assessing scholarly literature, which distinguishes it from conventional systematic literature reviews (De Bakker et al., 2005). The study by Köseoglu et al. (2016) entailed a quantitative examination of publication patterns within a certain academic discipline to evaluate the communication practices within that specific domain. Bibliometric analysis employs statistical computations to provide a quantitative evaluation of scholarly literature, comprising the description, appraisal, and monitoring of published research using bibliographic data. In this specific study, bibliographic data were acquired from Scopus, a globally recognized abstract and citation database of significant scale.

To identify the most recurrent research topics related to actions focusing on sustainable tourism that converge to increase the competitiveness of this important and strategic economic sector, we established a search on the Scopus Database, considering publications from 2000 and 2023.

The drafting and formalization of a protocol increase the objectivity of the information review process by offering a detailed description of the procedure developed to pose the research question, the study sample, the information obtained, and the exclusion criteria (Transfield et al., 2003). Table 3.1 depicts the protocol developed for this research.

The keywords used in this study were "competitiveness," "sustainability," and "tourism." A first searched made using the connector "OR" yielded 601 results, so we chose the connector "AND" between all terms in order to obtain 442 results.

Table 3.1 Narrative literature review protocol

Phase	Description
1	Defining the purpose of the literature review
2	Identification of the elements of the investigation
	Definition of the source of the information
	Definition of keywords and search terms
	Definition of the search period
3	Search and refinement
	Search Engine: Web of Science (WoS)
	Search by topic
	Use of specific keywords
	Selection of all years of the review period
4	Selection criteria
	Establishment of inclusion and exclusion criteria
5	Evaluation
	Review of title and abstract of articles
	Complete analysis

The search was filtered by language; 427 of the papers were written in English, and those in other languages were excluded. Subsequently, we performed filtering by area of knowledge ("business, management, and accounting," "social sciences," and "economics, econometrics and finance"), obtaining 357 documents. Thereafter, books, book chapters, reviews, and conference papers were excluded, obtaining 279 articles. Finally, we made a meticulous search excluding all keywords that were not related to the knowledge area, such as names of places or institutions and specific methodologies included in the papers. Eventually, we obtained a result of 115 documents.

Results

Annual Scientific Production on Competitiveness and Sustainability in Tourism

We recognize an increasing interest in the area since 2000, which has continued to the present. Now, we also observe a greater frequency of research works between 2014 and 2020, suggesting the impact of the COVID-19 pandemic on not only the global tourism sector but also related research.

Documents by Country

Interestingly, most of the academic research published on Scopus was on three European countries (Spain, Italy, and Portugal). The only Latin American country identified was Mexico with five documents. Indeed, the selection of the Scopus database and language implies the exclusion of other research initiatives; nonetheless, Latin American academics could do further research in this area.

Documents by Journals and Scholars

When observing the most cited journals, we noted the influence of sustainability as a global issue and the response of tourism as a related field, through the inclusion of titles related to ecology, environment, and climate change.

Moreover, iconic papers that are among the most cited, such as Buhalis et al. (2000), were identified. Marketing the competitive destination of the future. Tourism management, 21(1), 97–116 was cited 1708 times, and Scott, D., Gössling, S., and Hall, C. M. (2012). International tourism and climate change. Wiley Interdisciplinary Reviews: Climate Change, 3(3), 213–232 was cited 251 times (see Table 3.2). The journal Sustainability (Switzerland) had many cited articles, with 18 papers, and the most cited paper from the journal was Quaranta, G., Citro, E., and Salvia, R. (2016). Economic and social sustainable synergies to promote innovations in rural tourism and local development. Sustainability, 8(7), 668, cited 54 times (see Table 3.3).

Table 3.2 Most cited journals

Source title	SJR 2022	No. of articles	Cited by
Tourism Management	Q1	1	1,708
Wiley Interdisciplinary Reviews: Climate Change	Q1	1	251
Sustainability (Switzerland)	Q2	18	54
Journal of Sustainable Tourism	Q1	3	147
Sustainable Development	Q1	1	132
Current Issues in Tourism	Q1	5	53
Ecological Economics	Q1	1	80
Ecological Indicators	Q1	1	29
Journal of Travel Research	Q1	1	64
Journal of Vacation Marketing	Q1	1	51

Table 3.3 Most cited scholars

Most cited papers	Cited by	Citations per year	Keyword area			
			Economics	Competitiveness	Tourism	Sustainability
Buhalis, D. (2000)	1,708	74		X	X	
Scott, D., Hall, C. M., & Stefan, G. (2012).	332	30			X	X
Quaranta, G., Citro, E., & Salvia, R. (2016).	54	8		X	X	X
Scott, D., Gössling, S., & Hall, C. M. (2012).	251	23	X	X	X	
Scott, D., Simpson, M. C., & Sim, R. (2012).	147	13	X	X	X	
Streimikiene, D., Svagzdiene, B., Jasinskas, E., & Simanavicius, A. (2021).	132	66		X	X	
Alonso-Almeida, M. D. M., Bagur-Femenias, L., Llach, J., & Perramon, J. (2018).	53	11		X	X	X
Salvati, L., & Carlucci, M. (2011).	80	7	X	X	X	X
Armenski, T., Dwyer, L., & Pavluković, V. (2018).	64	13		X	X	
Lee, C. F., & King, B. (2009).	51	4		X	X	

Thematic Analysis

The first analysis has to do with the evolution of research topics over time. In this case, we can see how two large nodes are presented in the keywords that have been included in the 115 documents we finally selected. The areas of sustainability and economic growth stand out, as they are major research topics; it is interesting to point out how the paradigm shift from orthodox economic growth to sustainability represents a global challenge.

We can find competitiveness as a linking node between all categories: sustainability and tourism (which were part of the search) and economic growth and corporate social responsibility as resulting nodes on the keyword analysis. In contrast, we can see how economic growth, although it is a recurrent keyword in the field research works, does not hold a strong connection with the rest of the categories. The strongest links related to economic growth are categories well accepted in the orthodox economic theory such as fiscal policy, income distribution, economic impact, gross domestic product, exchange rate, and economic development. Nonetheless, the most cited papers in the search results were mainly related to environmental economy instead of economic growth.

There is a different situation related to sustainability, where the corresponding node holds strong links with not only competitiveness and tourism but also keywords related to business, corporate social responsibility, stakeholders' engagement, sustainable transition, sharing economy, sustainable tourism, and governance approach, among other issues that have been covered in recent research.

Finally, sustainability's link with the emerging research topics is mediated by tourism, and there are links between tourism and tourism behavior, tourist management, tourism development, gender, ecotourism, and destination competitiveness.

Discussion and Conclusion

Tourism, as can be seen in the results obtained, continues and will continue to be a strategic economic activity worldwide. It is a sector that generates economic growth in the communities where it is developed. This is in line with Guaita et al. (2020), who point out that tourism produces economic prosperity in tourist regions.

Competitiveness in the tourism sector still maintains the economic approach that gave rise to the concept, and it is important to keep in mind the factors that have an impact on said competitiveness. Factors related to strategic management, finance, innovation, and marketing, among others, are also important, in accordance with Mira et al. (2016), Salinas et al. (2022), and Zadeh and Kiliç (2021), who mention in their studies different business-related factors that influence the competitiveness of companies in the tourism sector. The COVID-19 contingency presented new challenges in this sector, and there are new approaches in accordance with global trends such as the United Nations 2030 Agenda for tourism and tourism competitiveness, as mentioned in Salinas et al. (2022), which identified the severe impact of the pandemic on the tourism industry as well as the world economy.

The approach to business competitiveness in the tourism sector has as perspectives environmental aspects that were not previously considered. The concept of competitiveness, which originated and evolved from the neoliberal economy, now focuses on the ecological economy, as Manniche et al. (2021) point out; the author includes issues related to sustainability within the economy.

This field of knowledge has experienced a significant increase in terms of academic production, mainly in relation to sustainability issues. This may be appreciated based on different aspects. First, the presence of economic growth as a significant node in the analysis is evident; however, the most recent research topics are shifting toward heterodox approaches of economics such as ecological economics; this can be confirmed based on the most cited journal titles. Among sustainability issues, other global concerns such as climate change, financial sustainability, sharing economy, and sustainable supply chain management appear. Tourism appears as a strong link between these emerging topics and concerns and the nodes of competitiveness and sustainability; meanwhile, competitiveness as a field of study has been displaced by sustainability, based on the migration to different approaches in economics and business studies. Therefore, along with the search of applicable studies, the tourism sector may continue to increase its research, especially into social and development problems, territory issues, sustainable goals, and competitiveness as a response to an increasingly complex world.

While it is important to acknowledge several limitations of this study, such as the utilization of a single database, the exclusion of alternative document formats, and the exclusion of non-English documents, the study's meticulous selection approach and use of a renowned database substantiate its significance as a point of reference for future researchers. To augment future research endeavors, it is advised to incorporate several databases, encompassing a wider range of documents of different types and languages. This approach would contribute to a more thorough understanding of the global research landscape pertaining to competitiveness, sustainability, and tourism.

References

Abreu-Novais, M., Ruhanen, L., & Arcodia, C. (2018). Destination competitiveness: A phenomenographic study. *Tourism Management, 64*(1), 324–334. https://doi.org/10.1016/j.tourman.2017.08.014

Alonso-Almeida, M., Llach, J., & Marimon, F. (2018). A Closer Look at the 'Global Reporting Initiative' Sustainability Reporting as a Tool to Implement Environmental and Social Policies: A Worldwide Sector Analysis. *Corporate Social Responsibility and Environmental Management, 21*(6), 318–335. Doi: https://doi.org/10.1002/csr.1318.

Alonso, S., Torrejón, M., Medina, M., & González, R. (2023). Sustainability as a building block for tourism – future research: Tourism agenda 2030. *Tourism Review, 78*(2), 461–474. https://doi.org/10.1108/TR-12-2021-0568.

Alves, S., & Ramos, A. (2015). Towards a sustainable tourism competitiveness measurement model for municipalities: Brazilian empirical evidence. *Revista de Tursimo y Patrimonio Cultural, 13*(6), 1337–1353. https://doi.org/10.25145/j.pasos.2015.13.093.

Armenski, T., Dwyer, L., & Pavluković, V. (2018). Destination Competitiveness: Public and Private Sector Tourism Management in Serbia. *Journal of Travel Research*, *57*(3), 384–398. Doi: https://doi.org/10.1177/0047287517692445.

Buhalis, D., Fan, D., Leung, X., & Darcy, S. (2023). Editorial: Tourism 2030 and the contribution to the sustainable development goals: The tourism review point. *Tourism Review*, *78*(2), 293–313. https://doi.org/10.1108/TR-04-2023-620.

Camisón, C. (2020). Competitiveness and sustainability in tourist firms and destinations. *Sustainability*, *12*(1), 1–6. https://doi.org/10.3390/su12062388.

Castro, U., González, J., & Maldonado, L. (2020). Economía Ecológica: Elementos para el Fomento del Turismo y la Sustentabilidad en Comunidades Rurales. In D. Gómez & M. Velarde (Eds.), *Turismo y Sustentabilidad en el Ámbito Rural* (pp. 11–36). Ediciones Navarra.

Chaigneau, T., Coulthard, S., Daw, T., Szaboova, L., Camfield, L., Stuart Chapin, F., Des Gasper, G., Hicks, C., & Ibrahim, M. (2022). Reconciling well-being and resilience for sustainable development. *Nature Sustainability*, *5*(4), 287–293. https://doi.org/41893-021-00790-8

Cheng, Y., Zhu, K., Zhou, Q., El Archi, Y., Kabil, M., Remenyic, B., & Dénes, L. (2023). Tourism ecological efficiency and sustainable development in the Hanjiang River Basin: A super-efficiency slacks-based measure model study. *Sustainability*, *15*(1), 1–17. https://doi.org/10.3390/su15076159.

Chong, K., & Balasingam, A. (2018). Tourism sustainability: Economic benefits and strategies for preservation and conservation of heritage sites in South East Asia. *Tourism Review*, *74*(2), 268–279. https://doi.org/10.1108/TR-11-2017-0182.

Correa, F. (2006). Antecedentes y Evolución de la Economía Ecológica. *Revista Semestre Económico*, *9*(17), 13–41.

Costanza, R., Erickson, J., Farley, J., & Kubiszewski, I. (2020). *Sustainable wellbeing futures: A research and action agenda for ecological economics*. Edward Elgar Publishing.

De Bakker, F., Groenewegen, P., & Den Hond, F. (2005). A bibliometric analysis of 30 years of research and theory on corporate social responsibility and corporate social performance. *Business & Society*, *44*(3), 283–317.

De Castro, M., Fernández, P., Guaita, J., & Martín, J. (2020). Modelling natural capital: A proposal for a mixed multi-criteria approach to assign management priorities to ecosystem services. *Contemporary Economics*, *14*(1), 22–37. https://doi.org/10.5709/ce.1897-9254.330.

Elmo, G., Arcese, G., Valeri, M., Poponi, S., & Pacchera, F. (2020). Sustainability in tourism as an innovation driver: An analysis of family business reality. *Sustainability*, *12*(15), 31–58. https://doi.org/10.3390/su12156149.

Fakfare, P., & Wattanacharoensil, W. (2021). Impacts of community market development on the residents' well-being and satisfaction. *Tourism Review*, *76*(5), 1123–1140. https://doi.org/10.1108/TR-02-2020-0071.

Falatoonitoosi, E., Schaffer, V., & Kerr, D. (2022). Does sustainable tourism development enhance destination prosperity? *Journal of Hospitality & Tourism Research*, *46*(5), 1056–1082. https://doi.org/10.1177/1096348020988328.

Fernandez, J., & Rivero, M. (2009). Measuring tourism sustainability: Proposal for a composite index. *Tourism Economics*, *15*(2), 277–296.

García, M. (2003). Apuntes de Economía Ecológica. *Boletín económico de ICE*, (2767), 69–75.

Gross, S., & Grimm, B. (2018). Sustainable mode of transport choices at the destination – public transport at German destinations. *Tourism Review*, *73*(3), 401–420. https://doi.org/10.1108/TR-11-2017-0177.

Guaita, J., Martín, J., & Salinas, J. (2020). Innovation in the measurement of tourism competitiveness. In M. Galindo, M. Méndez, & M. Castaño (Eds.), *Analyzing the relationship between innovation, value creation and entrepreneurship* (pp. 268–288), IGI-Global. https://doi.org/10.4018/978-1-7998-1169-5.ch013.

Köseoglu, M., Rahimi, R., Okumus, F., & Liu, J. (2016). Bibliometric studies in tourism. *Annals of Tourism Research, 61*(5), 180–198. https://doi.org/10.1016/j.annals.2016.10.006

Leach, M., Raworth, K., & Rockström, J. (2013). *Between social and planetary boundaries: Navigating pathways in the safe and just space for humanity.* OCDE.

Lee, C.-F., & King, B. (2009). A Determination of Destination Competitiveness for Taiwan's Hot Springs Tourism Sector using the Delphi Technique. *Journal of Vacation Marketing, 15*(3), 243–257. Doi: https://doi.org/10.1177/1356766709104270.

Lew, A. (2020). The global consciousness path to sustainable tourism: A perspective paper. *Tourism Review, 75*(1), 69–75. https://doi.org/10.1108/TR-07-2019-0291.

Manniche, J., Larsen, K., & Broegaard, R. (2021). The circular economy in tourism: Transition perspectives for business and research. *Scandinavian Journal of Hospitality and Tourism, 21*(3), 247–264. https://doi.org/10.1080/15022250.2021.1921020.

Mira, M., Moura, A., & Breda, Z. (2016). Destination competitiveness and competitiveness indicators: Illustration of the Portuguese case. *Review of Applied Management Studies, 14*, 90–103. http://dx.doi.org/10.1016/j.tekhne.2016.06.002.

Morgan, J. (2017). Piketty and the growth dilemma revisited in the context of ecological economics. *Ecological Economics, 136*(1), 169–177. https://doi.org/10.1016/j.ecolecon.2017.02.024.

Nunkoo, R., Sharma, A., Rana, N., Dwivedi, Y., & Sunnassee, V. (2021). Advancing sustainable development goals through interdisciplinarity in sustainable tourism research. *Journal of Sustainable Tourism, 31*, 1–25. https://doi.org/10.1080/09669582.2021.2004416.

Pulido-Fernandez, J., Cardenas-Garcia, P., & Espinosa-Pulido, J. (2019). Does environmental sustainability contribute to tourism growth? An analysis at the country level. *Journal of Cleaner Production, 213*, 309–319. https://doi.org/10.1016/j.jclepro.2018.12.151.

Quaranta, G., Citro, E., & Salvia, R. (2016). Economic and Social Sustainable Synergies to Promote Innovations in Rural Tourism and Local Development. *Sustainability, 8*, 668. Doi: https://doi.org/10.3390/su8070668.

Ritchie, J., & Crouch, G. (2010). A model of destination competitiveness/sustainability: Brazilian perspectives. *Revista de Administración Pública, 44*(5), 1049–1066. https://doi.org/10.1590/S0034-76122010000500003.

Salinas, J., Guaita, J., & Martin, J. (2022). An analysis of the competitiveness of the tourism industry in a context of economic recovery following the COVID-19 pandemic. *Technological Forecasting & Social Change, 174*(1), 123–146. https://doi.org/10.1016/j.techfore.2021.121301.

Scott, D., Gössling, S., & Hall, C. (2012). International Tourism and Climate Change. *Advanced Review, 3*(3), 213–232. Doi: https://doi.org/10.1002/wcc.165.

Sigala, M. (2020). Tourism and COVID-19: Impacts and implications for advancing and resetting industry and research. *Journal of Business Research, 117*(1), 312–321. https://doi.org/10.1016/j.jbusres.2020.06.015.

Streimikiene, D., Svagzdiene, B., Jasinskas, E., & Simanavicius, A. (2021). Sustainable tourism development and competitiveness: The systematic literature review. *Sustainable Development, 29*(1), 259–271. https://doi.org/10.1002/sd.2133.

Tasci, A. (2017). Consumer demand for sustainability benchmarks in tourism and hospitality. *Tourism Review, 72*(4), 375–391. https://doi.org/10.1108/TR-05-2017-0087.

Tasci, A., Fyall, A., & Woosnam, K. (2021). Sustainable tourism consumer: Sociodemographic, psychographic and behavioral characteristics. *Tourism Review, 77*(2), 341–375. https://doi.org/10.1108/TR-09–2020–0435.

Transfield, D., Denyer, D., & Smart, P. (2003). Towards a methodology for developing evidence-informed management knowledge by means of systematic review. *British Journal of Management, 14*(3), 207–222. https://doi.org/10.2307/249689.

Valeri, M. (2015). Sustainability development and competitiveness of Rome as a tourist destination. *Tourism and Hospitality Management, 21*(2), 203–217.

Valeri, M. (2023). *Tourism innovation in the digital era. Big data, AI and technological transformation.* Emerald Publishing, UK.

Valeri, M. (2024). *Innovation strategies and organizational culture in tourism. Concepts and case studies on knowledge sharing.* Routledge Publishing.

Weaver, D. (2011). Organic, incremental and induced paths to sustainable mass tourism convergence. *Tourism Management, 33*(5), 1030–1037.

World Tourism Organization. (2017). *The tourism sector highlights the potential of urban tourism and the need to move toward more sustainable practices.* Retrieved August 29, 2023, from www.unwto.org/archive/global/press-release/2017-05-12/tourism-sector-highlights-poten-tial-urban-tourism-and-need-move-toward-more.

Yang, Y., Yang, H., & Cheng, Y. (2021). Why is it crucial to evaluate the fairness of natural capital consumption in urban agglomerations in terms of ecosystem services and economic contribution? *Sustainable Horizons, 4*(1), 1–35. https://doi.org/10.1016/j.horiz.2022.100035.

Zadeh, R., & Kiliç, H. (2021). Tourism competitiveness and tourism sector performance: Empirical insights from new data. *Journal of Hospitality and Tourism Management, 46*(1), 73–82. https://doi.org/10.1016/j.jhtm.2020.11.011.

4 The New Economic Paradigm Formed by Metaverse in the Tourism Sector

Touristic GIG Economy

Sabri Öz and Pınar Başar

Introduction

Technological and digital transformation is realizing the most important paradigm shifts of the 21st century. Especially after COVID-19, every component of digital transformation, disciplines such as data integration, big data, AI, and cyber security began to rapidly show their impact in every field. In this context, virtual reality and AR/VR technologies have also accelerated, especially with metaverse structures. It is seen that there are intensive practices to be effective in every field of the services sector.

With remote working after COVID-19 and the transformation of many jobs and studies to online platforms, the way of working has also caused radical transformations in labor markets. The transformation process is still ongoing. It is seen that AR/VR technologies have an impact in many areas, especially in the gaming industry. The most important working platforms in this field are associated with the metaverse.

The purpose of this study is to determine the weight of metaverse applications included in the GIG Economy in terms of PESTLE (Political, Economical, Social, Technological, Law-Legislation, and Environmental aspects) components. In other words, it is to investigate the answer to the question of what way metaverse applications will affect the tourism industry. The methodology of the study can be called mixed-method. The combination of literature review, bibliometric analysis, analytical hierarchical process (AHP) analysis, and PESTLE analysis seems to be in line with the mixed method approach.

The plan of the study includes a literature review in the first part, which will be related to tourism on a conceptual basis, and then a bibliometric analysis where all concepts are used together and carried out with the keywords of the study. In the second part, there is an application where AHP Analysis and PESTLE Analysis are evaluated together. The study will end with the Findings and Discussion section where the analysis results are evaluated.

Literature Review and Bibliometric Analysis

There are many intensive studies on tourism in the literature. There are over 123,000 studies on tourism on the Web of Science (WoS) platform alone. In this section,

DOI: 10.4324/9781003497004-5

studies containing the keywords GIG Economy, metaverse, and AHP along with tourism will be questioned.

GIG Economy and Tourism

Of the 123,000 studies mentioned earlier, while approximately 30,000 are related to technology, only 17 studies contain the words "GIG Economy" and "Tourism" together in the title, abstract, and keywords of the studies (TS=GIG Economy AND TS=Tourism) (Clarivate, 2023). In Figure 4.1, the 123,000 tourism-related studies on the WoS platform are categorized.

Figure 4.1 shows that there are a total of 30,963 studies in the fields of management, business, technology, and social sciences, assuming that they are completely or partially related to technology.

In studies involving the concepts of tourism and GIG Economy, studies on the sharing of income of the sharing economy can be found in the field of tourism (Chen et al., 2022). It can be seen that the GIG Economy is discussed with its positive and negative aspects (C. Kobis et al., 2021). Just as the resignations that came with the COVID-19 epidemic and are closely related to the tourism sector have been examined, it is seen that specific areas have also been studied and a relationship has been established with the GIG Economy (Liu-Lastres et al., 2023).

When GIG Economy is mentioned, important platforms such as Uber, Airbnb, and Amazon come to mind. In this context, studies related to the GIG Economy, sharing economy, and Airbnb are seen and the effects are analyzed (Cassell & Deutsch, 2023; Hall et al., 2022).

It is seen that there are also studies on the basis of countries and regions in the field of GIG Economy and tourism and, again, studies on Ashmania and Cuba where the Airbnb effect was investigated (Cassell & Deutsch, 2023; Nemer et al., 2018). Additionally, it is possible to see a study on Malaysia (Nga et al., 2021)

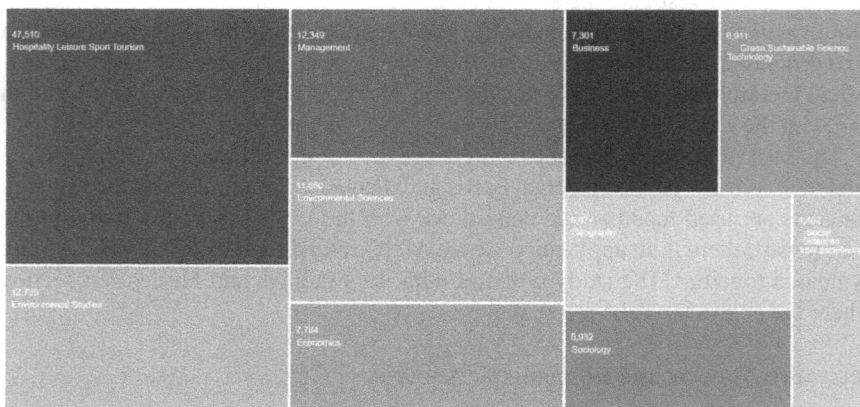

Figure 4.1 Tree map of WoS categories of the query: "TS=Tourism" (Clarivate, 2023).

including COVID-19 and unemployment and studies on the same subject in the Southern African region (Baum & Giddy, 2021).

There are also studies on bibliometric analysis of the GIG Economy, especially the sharing economy (Gürsoy, 2023; Klarin & Suseno, 2021). In addition to bibliometric studies, systematic scanning and mapping studies have also been seen (Laurenti et al., 2019).

Metaverse and Tourism

Although metaverse has started to gain popularity, especially with the ecosystem to be created on blockchain, its influence has started to decrease recently. However, with the spread and especially the development of AR/VR technologies and the beginning of the study of MR and XR disciplines, R&D activities carried out on a sectoral basis also become valuable. Just looking at the titles on the WoS platform, there are 23 studies containing the words tourism and metaverse (TI=Metaverse AND TI=Tourism).

When we look at the studies on countries in Figure 4.2, we see that the distribution is concentrated in certain regions.

According to Figure 4.2, seven of the 23 studies are on the People's Republic of China and four are on the USA, accounting for approximately half (48%) of the studies.

Since the subject of metaverse is still a new understanding, it is seen that conceptual studies are included and future approaches are discussed (Gursoy et al., 2022; Koo et al., 2022; Volchek & Brysch, 2023; F. X. X. Yang & Wang, 2023).

In studies on metaverse and tourism, some analyses have focused on concerns and anxieties (Buhalis, Leung, et al., 2023; Monaco & Sacchi, 2023; Wei, 2024).

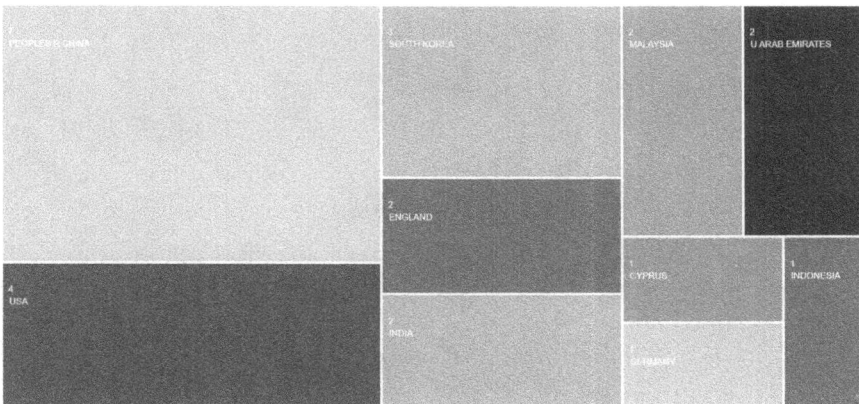

Figure 4.2 Tree map of countries/regions of the query: "TI=Tourism AND TI=Metaverse" (Clarivate, 2023).

Some of them include positive contributions to the tourism market and what positive results it will bring in the future (Tsai, 2022; Ucgun & Sahin, 2023).

In studies that directly express the relations between the tourism market and the metaverse, the effects and roles of other disciplines of technological and digital transformation have been investigated, and it has been observed that the most intense relationship in this field is created by artificial intelligence (AI) and blockchain (Jo, 2023; Lin et al., 2023).

In addition to these studies, it is seen that there are also studies in the literature on a sectoral basis (such as smart tourism, health field, sports field, and muse tours) (Buhalis, Lin, et al., 2023; Sanchez-Amboage et al., 2023; Shaygani et al., 2022; Suanpang et al., 2022).

AHP Analysis and Tourism

When looking at AHP studies in the field of tourism on the WoS platform, the number of documents found is 56. It is seen that more than half of the studies are related to the environment and sustainability, and very few of them are focused on technological and digital transformation. In a sense, this reveals the distinctive feature of this study. Figure 4.3 shows the distribution of AHP and tourism-related studies by WoS category.

As can be seen from Figure 4.3, while there are 36 studies on environment and sustainability, fields directly based on technology are much less.

When the studies are examined, it is seen that AHP is generally used with Fuzzy AHP or a different derivative (such as Fuzzy AHP) and the focus is on weighting. AHP has become a particularly preferred method for weighting the criteria that will affect decisions in multi-criteria decision-making techniques. It is considered an advanced tool for criterion weighting, especially in studies based on qualitative

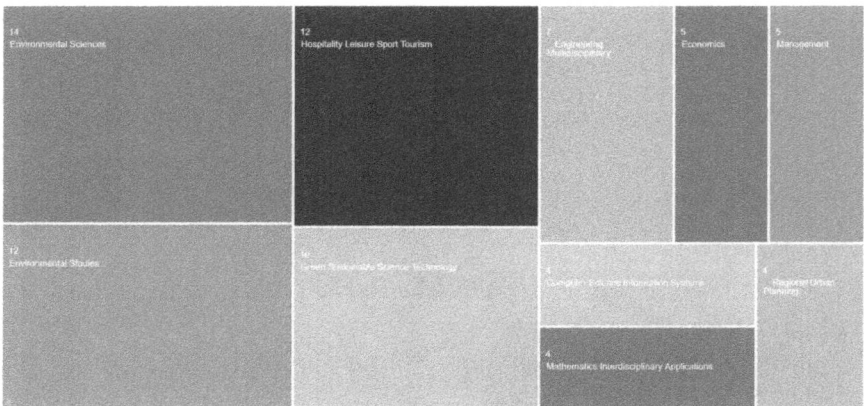

Figure 4.3 Tree map of WoS categories of the query: "TI=AHP AND TI=Tourism" (Clarivate, 2023).

and subjective data (Buyukozkan et al., 2021; Cao, 2022; Kim et al., 2017; Tian et al., 2020; Wang et al., 2016; S. Yang & Kong, 2022).

Criteria weighted by the AHP method are important in scoring alternatives and providing input in choosing among alternatives. When we look at the studies on the tourism sector using multi-criteria decision-making techniques in choosing among alternatives, it can be seen that there is no intensive work on the WoS platform. It is seen that Topsis and Promethee methods are used in choosing between alternatives (Butowski, 2018; Tian et al., 2020; Ulkhaq et al., 2016).

Combining the Scope via Bibliometric Analysis

When working on the WoS platform with an overview of the keywords of this study, the following results are obtained:

- TI=(Tourism) AND (TS=("GIG Economy") OR TS=(Metaverse) OR TS= (AHP)) gives 274 results whereas
- TI=(Tourism) OR TI=("GIG Economy") OR TI=(Metaverse) 61,857 and
- When we narrow the research for the last five years (2020-2021-2022-2023-2024), it is seen that 17,985 documents were obtained.

In this section, the bibliometric analysis of the 274 studies obtained, and subsequently, the 17,985 data will be briefly reviewed with the VosViewer program.

In the study, where 274 printouts were taken, a search was made with the keywords Metaverse, AHP, and GIG economy, which are the subject of this study, depending on the tourism keyword. The data were exported as input to the Vosviewer program and then the program was run. Based on 274 studies, a total of 857 keywords were found and 27 of them were repeated at least five times. Figure 4.4 shows the distribution and relationships of keywords.

Overlay visualization is a network visual that expresses how often keywords are used over time and how often they are used in relation to which keywords. It is stated in Figure 4.4 that tourism and AHP and derivative keywords have been intensively studied in the past (2014–2018 periods), whereas, as a part of the GIG Economy, topics related to metaverse, VR, and sustainability have been intensively studied recently. When looking at the study density (the size of the spheres symbolizing the keywords), it can be interpreted that recent studies (Metaverse and VR) can be studied more. When the link and clustered keywords are examined, it is seen that these three separate disciplines are strong between AHP and tourism, but weaker with the fields representing the GIG Economy. This is why this study is important.

Figure 4.5 shows the clustered keywords.

In Figure 4.5, the aforementioned separation can be seen more clearly. Keywords related to the GIG Economy appear independently of the others in Cluster 3.

On the other hand, when 17,985 of the 61,857 documents obtained with the (TI=Tourism OR TI=GIG Economy OR TI=Metaverse) SQL query in the last five

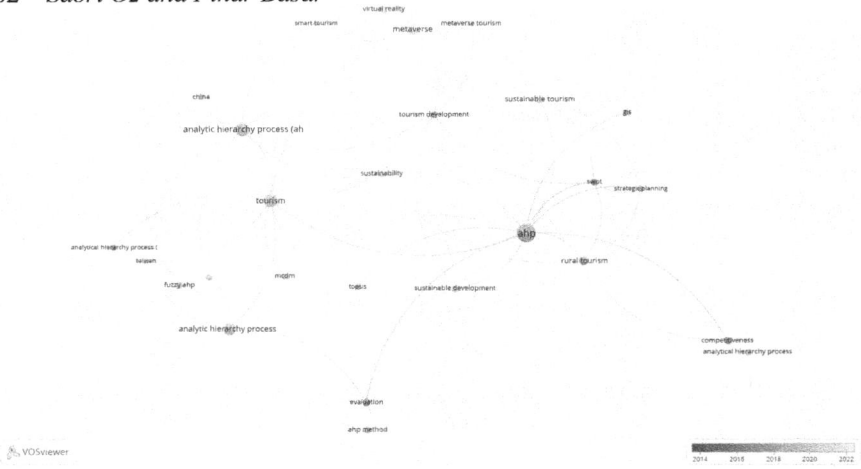

Figure 4.4 Overlay visualization of the query on WoS: TI=Tourism AND (TS=GIG Economy OR TS=Metaverse OR TS=AHP) (Vosviewer, 2023).

27 items (6 clusters):

Cluster 1 (8 items)
analytic hierarchy process
analytical hierarchy proces
fuzzy ahp
mcdm
medical tourism
taiwan
tops1s
tourism

Cluster 3 (5 items)
metaverse
metaverse tourism
smart tourism
tourism development
virtual reality

Cluster 2 (6 items)
ahp
gis
rural tourism
strategic planning
sustainable tourism
swot

Cluster 4 (3 items)
ahp method
evaluation
sustainable development

Cluster 5 (3 items)
analytic hierarchy process
china
sustainability

Cluster 6 (2 items)
analytical hierarchy proces
competitiveness

Figure 4.5 Clusters of the query on WoS: TI=Tourism AND (TS=GIG Economy OR TS=Metaverse OR TS=AHP) (Vosviewer, 2023).

years are examined in their entirety, the results of the WoS platform on the work areas are shown in Figure 4.6.

Figure 4.6 shows that of the 17,985 studies that included the words tourism, GIG Economy, or metaverse in their study titles, more than half focused on the environment, sustainability, and types of tourism.

When 17,985 documents are studied in the VosViewer program, the overlay visualization output of the keywords is seen in Figure 4.7.

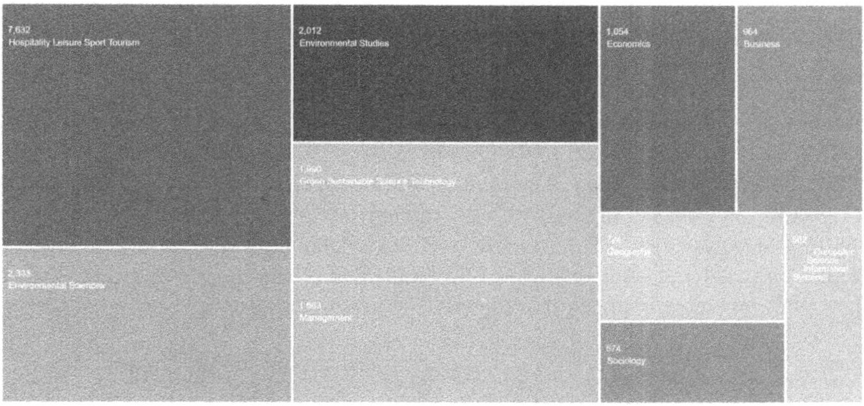

Figure 4.6 Tree map of WoS categories of the query: TI=Tourism OR TI=GIG Economy OR TI=Metaverse (Clarivate, 2023).

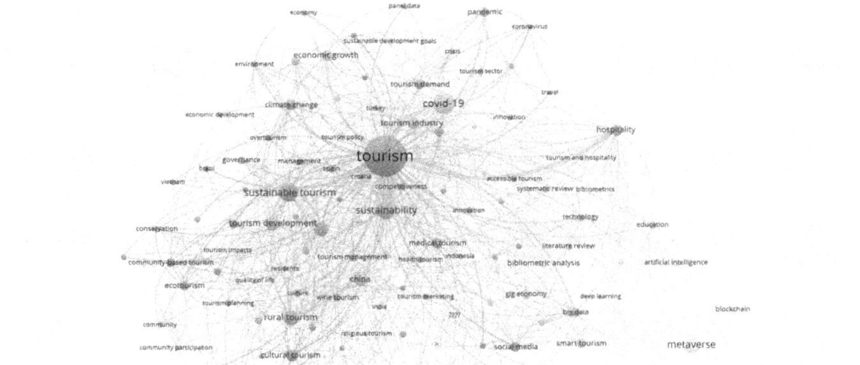

Figure 4.7 Overlay visualization of the query on WoS: TI=Tourism OR TI=GIG Economy OR TI=Metaverse (Vosviewer, 2023).

Interestingly, when looking at the picture from above, areas that include technological and digital transformation such as VR, metaverse, blockchain, and AI, which can be considered parts of the GIG Economy, are both less studied and less studied than keywords such as tourism, sustainability, COVID-19 in previous years. It can be seen in Figure 4.7 that it has been studied. In the analysis of 17,985 documents, it was found that 33,237 different keywords were used and 125 keywords shown in Figure 4.7 were repeated at least 50 times (Vosviewer, 2023).

The density chart of 125 keywords for the same query, with the same settings, regardless of time, shows that the GIG Economy is not sufficiently associated with tourism, regardless of time, and is still studied at low intensity.

Methodology: AHP Analysis via Pestle Components

In this section, the implementation of the study in the field is seen. The study gathered demographic information of the business people and conducted interviews with nine experienced business people on tourism, technological, and digital transformation, and PESTLE analysis was carried out in accordance with the AHP analysis structure. PESTLE analysis studied which political, economic, social, technological, legal, and environmental dimensions are more important for the future of metaverse, representing the GIG Economy in tourism.

Contributors to AHP Analysis

In this study, evaluation was made with the fuzzy logic version of AHP analysis. Demographic information of the nine sector representatives participating in the research is shown in Table 4.1.

Participants in the business world, who have been familiar with the tourism sector for at least 15 years, were encouraged to think about the future of the tourism sector, and especially by giving preliminary information about the GIG Economy, the purpose was explained and their opinions were received through a

Table 4.1 Demographic data of participants

Name/surname	Experience	Field	Sector	Location	Code
Ö**K**	42	Management	Aviation	Italy	P1
M**Ö**	16	Service	Tourism	Misc.	P2
V**Ü**	31	Catering Owner	Tourism	Türkiye	P3
Ö**F**Y**	18	Hotel. Owner	Service	Türkiye	P4
R**Y**	16	Manager	IT-Tourism	Hungary	P5
P**Ö**B**	19	CEO	Restaurant	Germany	P6
S**Y**O**	20	Co. Owner	Service	Türkiye	P7
V**F**B**	20	Manager	Tourism Finance	Hungary	P8
S**H**	25	Management	Aviation-Cargo	Misc.	P9

semi-structured interview in accordance with the AHP rules. In the tourism sector, the structures on the metaverse were explained, examples were given, and then the aspects in which the tourism sector would be affected were emphasized.

The first question will be,

> Considering the impact of sightseeing tours on the Tourism sector in virtual environments such as Metaverse, is the "Political" impact or the "Economic" impact more important? If you were to rate the level of importance, would you say how many times it is effective, with 1 being the least effective and 9 being the most effective?

The same question frame would be asked by changing the impact factors such as political, economic, social, technological, legislation, and environmental, which are the components of PESTLE.

Particular efforts were made to ensure that the business people interviewed had the same issue to understand with each question. Although the participants coded P1, P3, P8, and P9 emphasized that the issue would vary depending on the sector, especially when evaluating the political and legal situation, they were persuaded to make an average judgment.

AHP Analysis

While performing the AHP Analysis, the six criteria considered were asked to be compared in pairs, and a comparison was requested based on numerical data regarding which criterion was more important and its degree of importance. In this analysis method developed by Saaty (SAATY, 1966; Saaty, 1974, p. 2) in order to minimize subjectivity, information was given in the field of GIG Economy and metaverse and discussions were made about the applications. Although none of the participants had metaverse experience, it was observed that they were knowledge-able about metaverse and AR/VR technologies.

A total of 30 questions were asked to each participant, as they were asked to make pairwise comparisons of six criteria and then rate them. (There are two questions in each comparison – which one is important and how important it is.) A minimum of 72 and a maximum of 91 minutes of study was conducted with each participant.

The information obtained from the studies was compiled using the MS Excel program for AHP analysis to create a decision matrix. For each comparison, the value from each participant was reduced to a single value to create its geometric mean. For example, when each participant was asked which of the economic and environmental factors was more important, the values given in Table 4.2 were obtained, and the calculation is also shown in Table 4.2.

As shown in Table 4.2, P3 said that the environmental issue is more important than economic criteria when thought the impact of metaverse on tourism. P4 said that both are of equal importance. And the rest of the participants thought that the economic factor is more important than the environmental factor. Therefore, the

Table 4.2 Comparison of economic (Ec) and environmental (En) criteria for metaverse on tourism

	P1	P2	P3	P4	P5	P6	P7	P8	P9
Which one is important?	Ec	Ec	**En**	=	Ec	Ec	Ec	Ec	Ec
The level of importance?	2	3	3	1	3	2	3	2	3

Geometric average = 9th root of: 2*3*1/3*1*3*2*3*2*3 =1.82
(Economic)

Initial Matrix

	P	Ec	S	T	L	En
P	1.000	0.183	0.127	0.121	0.189	0.149
Ec	5.475	1.000	2.961	0.333	3.769	1.817
S	7.882	0.338	1.000	0.523	1.489	0.914
T	8.258	3.000	1.912	1.000	3.425	4.215
L	5.294	0.265	0.672	0.292	1.000	0.783
En	6.730	0.550	1.094	0.237	1.277	1.000
Sum	34.639	5.336	7.766	2.507	11.149	8.878

Normalized Matrix

	P	Ec	S	T	L	En	weight
P	0.029	0.034	0.016	0.048	0.017	0.017	0.027
Ec	0.158	0.187	0.381	0.133	0.338	0.205	0.234
S	0.228	0.063	0.129	0.209	0.134	0.103	0.144
T	0.238	0.562	0.246	0.399	0.307	0.475	0.371
L	0.153	0.050	0.086	0.116	0.090	0.088	0.097
En	0.194	0.103	0.141	0.095	0.115	0.113	0.127
Sum	1.000	1.000	1.000	1.000	1.000	1.000	

Weighted Matrix

	P	Ec	S	T	L	En	sum	sum/weight
P	0.027	0.043	0.018	0.045	0.018	0.019	0.170	6.320
Ec	0.147	0.234	0.427	0.124	0.367	0.230	1.528	6.538
S	0.212	0.079	0.144	0.194	0.145	0.116	0.890	6.174
T	0.222	0.701	0.276	0.371	0.333	0.534	2.437	6.564
L	0.142	0.062	0.097	0.108	0.097	0.099	0.606	6.233
En	0.181	0.129	0.158	0.088	0.124	0.127	0.806	6.365

Eigenvalue. e. (Average of the sum/weight) :   6.366
Consitancy Index. CI. [CI = (e−6)/(6−1)] :   0.073
Consistancy Ratio. CR. [Consitant if CR = CI/R16 < 0.10] :   0.058

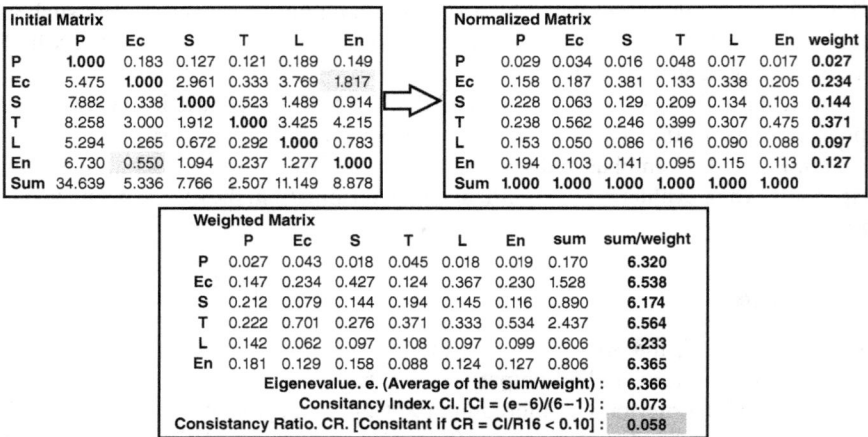

Figure 4.8 Calculations for AHP analysis.

geometric average of economic factor is 1.82 times more important than the environmental issue when it is thought the impact of metaverse on tourism.

As Figure 4.8 depicts, since economic criteria are 1.82 times more important than the environment, the opposite is also relevant: Environmental issue is 1/1.82 = 0.55 times more important than economic issue.

Table 4.3 shows the AHP output.

As seen in Table 4.3, it is stated that metaverse and GIG Economy will have the most impact on the tourism sector with its technological aspect. In this context, it can be interpreted that the tourism sector should develop a strategy by focusing on the technological field. The economic criterion comes second. In this context, investment and budgeting will be of great importance. The fact that the social dimension of the issue should not be ignored comes third.

Findings and Discussion

The most intense criticism of AHP analysis stems from the fact that it starts with subjective evaluations. In this study, the fact that each participant was informed

Table 4.3 Output of the AHP

Number of parameters (components), n		6
Calculated eigenvalue, e		6.375
Consistency index, CI = (e-n)/(n-1)		0.075
Random Saaty index, RI_6		1.252
Consistency ratio, CR=CI/RI_6		0.06
Analysis result (CR<0.10 is consistent)		**Consistent**
Components	**Weight**	**Rank**
(P)olitical	0.03	6
(Ec)onomical	0.24	2
(S)ocial	0.14	3
(T)echnological	0.37	1
(L)egislation and laws	0.10	5
(En)vironmental	0.13	4

to ensure a homogeneous understanding of the parable, was informed about the study in advance, and provided the participants with the opportunity to prepare was intended to alleviate this criticism.

On the other hand, would the situation be different if the opinions received from the participants were evaluated with direct questions without any information or even a conversation? This is a matter of debate, but the method applied in this study was informed.

As stated in the sections above, some participants emphasized that the answers given to the questions may vary depending on the type of tourism. Therefore, another issue that can be discussed is "what would the result be if the study was developed on a sectoral scale?" may be the answer to the question. Therefore, it may be useful to renew this study for different species in different tourism fields. Different results may occur in health tourism, beach tourism, educational tourism, or sports tourism, and important results can be obtained in developing strategies according to the types of tourism.

In this study, metaverse and AR/VR technologies, as a component of the GIG Economy, were scanned through a bibliometric study, and analyses were made based on this. The studies carried out can also be supported by studies carried out in other countries. Different conclusions can also be drawn from the fact that studies on countries and regions are concentrated.

It has long been a matter of debate why bibliometric analysis is performed only on WoS. WoS is one of the leading database hosting platforms in the world. There may be sectoral advantages in platforms from time to time. In the field of social sciences, WoS is considered sufficient. Studies have shown that, for example, a bibliometric analysis conducted on Scopus and the WoS platform gives similar results (Öz, 2023).

As can be seen from the literature review and bibliometric analysis, there are other components of the GIG Economy. When metaverse and other components are compared, a study can be conducted to determine which one will have a more intense impact on tourism. While this is a matter of discussion for this study, it is also important in terms of guiding other studies. Components and disciplines involved in technological and digital transformation, such as AI, blockchain, or cyber security, may produce different results. Moreover, these components are disciplines that also play an active role in the GIG Economy.

The consistency of the results and opinions obtained with AHP analysis, in a sense, represents a known or predicted fact. In this study, this situation can be considered proven with an analytical methodology. Essentially, depending on this, strategic road maps to be implemented by the sector can be shaped. If technology is the first important factor in the analysis handled with AHP, what could be the next step to evaluate this result? Since AHP provides an input into the algorithmic methodologies to be used in choosing between alternatives, how can it guide a tourism company that wants to improve its technology? As an answer to this question, one can try to consider the components of Industry 4.0 as alternatives and choose between them. By using methods such as Topsis or Promethee, Electra, a study can be produced that includes the AHP output and explains and guides the investor on which component should be used in technological and digital transformation.

In a mechanism where alternatives are ready, criticism based on subjective opinions on the AHP side can be eliminated by using the CRITIC method, one of the multi-criteria decision-making techniques.

Conclusion

Technological and digital transformation is very effective in the tourism sector, as in every field. Tourism is the most heavily affected and damaged sector due to COVID-19. It has to develop actions for both competitive advantage and efficient and effective use of resources through technological transformation. This being the case, the GIG Economy, which emerged with the influence of technological developments, causes new paradigms to emerge. In the GIG Economy, metaverse applications are the most important component. People can perform many actions such as traveling to other universes from where they are, creating other universes, and even visiting museums and participating in sports activities and events. Moreover, they have the opportunity to carry out these activities by actually touching, feeling, and experiencing the three-dimensional designs at a lower cost, right from their home, without wasting time, without changing places, and without experiencing transportation problems. Therefore, metaverse was deemed worthy of research for tourism within the GIG Economy, and what its effects could be and with what criteria it could affect the sector were investigated.

In the research, a detailed bibliometric analysis was conducted on the GIG Economy, AHP analysis, tourism, and metaverse. According to the analysis made on the WoS platform, it seems that there are very few studies on the metaverse and the tourism sector, that studies have started recently, and that it is still an open subject to be studied. Bibliometric analysis also shows that AHP analysis is supported by intensive studies accepted in the tourism sector. Despite the critical approaches to AHP, it has been observed that it is also used extensively in the tourism sector in weighting criteria in terms of multi-criteria decision-making techniques. For this reason, this study makes this study unique in that it is a study in which GIG Economics and AHP methodology have been applied together in the tourism sector, and its intensive use means that AHP can be applied to tourism.

The originality of the study should be stated once again, as a study that assumes the PESTLE components of AHP analysis as criteria and applies these criteria on the GIG Economy (and therefore metaverse) and tourism. In this original study, the Political, Economic, Social, Technological, Legal, and Environmental effects, which are among the PESTLE components, were accepted as the effects of metaverse applications on tourism, and these criteria were weighted with AHP analysis. As a result of the study and analysis, it was seen that technology was the most important criterion, as stated by nine participants. It is accepted that technological factors have a very important place in establishing competitive advantages, maintaining and developing competitive advantages, resource use, and efficiency. In this regard, the technological factor is the most important factor (0.37), which carries a positive message about the future, the planning of businesses, the investments they will make in strategic areas, and the roadmaps they will follow toward digitalization. In this case, it can be suggested that they should accelerate their work on alternatives to technological and digital transformation.

The second and most important criterion is seen as the economic factor (0.24). It can be thought that the reason why the economic factor is so important is because technology, especially AR/VR technologies, is already a capital-intensive product. For this reason, this study emphasizes that investments in technological and digital transformation are necessary and that budgeting and financing are very important.

The third important factor appears to be the social factor. According to the results of the AHP analysis, this factor, which has a weight of 0.14, is one of the most discussed areas for metaverse in technological and digital transformation, but tourism is more positive in this sense.

Environmental factors have been discussed very intensively recently in areas such as climate crisis and green energy. Tourism has long been referred to as "industry without smokes." It means that it is chimney-free, therefore smoke-free, and therefore has the lowest level of greenhouse gases. Traffic and transportation are the most important environmental pollutants in the tourism sector. Since an application like metaverse eliminates the transportation issue, the harmful factors will also be eliminated. For this reason, it is normal that it ranks fourth in the evaluation.

There is no applicable legal regulation regarding metaverse platforms yet. In countries that have norms in certain areas, it is not possible for it to work with all its content because new applications are developed every day. For these and similar reasons, both state legal regulations and policy criteria have been determined as factors at the lowest level.

References

Baum, T., & Giddy, J. K. (2021). The gig economy and employment in tourism in Southern Africa A global finger in the informal sector pie? In J. Saarinen & J. M. Rogerson (Eds.), *Tourism, change and the global south* (pp. 222–236). Routledge. www.webofscience.com/wos/woscc/summary/66917c60-3a59-488f-ac34-2801772b5810-a4420f75/relevance/1

Buhalis, D., Leung, D., & Lin, M. (2023). Metaverse as a disruptive technology revolutionising tourism management and marketing. *Tourism Management, 97,* 104724. https://doi.org/10.1016/j.tourman.2023.104724

Buhalis, D., Lin, M. S., & Leung, D. (2023). Metaverse as a driver for customer experience and value co-creation: Implications for hospitality and tourism management and marketing. *International Journal of Contemporary Hospitality Management, 35*(2), 701–716. https://doi.org/10.1108/IJCHM-05-2022-0631

Butowski, L. (2018). An integrated AHP and PROMETHEE approach to the evaluation of the attractiveness of European maritime areas for sailing tourism. *Moravian Geographical Reports, 26*(2), 135–148. https://doi.org/10.2478/mgr-2018-0011

Buyukozkan, G., Mukul, E., & Kongar, E. (2021). Health tourism strategy selection via SWOT analysis and integrated hesitant fuzzy linguistic AHP-MABAC approach. *Socio-economic Planning Sciences, 74,* 100929. https://doi.org/10.1016/j.seps.2020.100929

Cao, S. (2022). Development potential evaluation for land resources of forest tourism based on fuzzy AHP method. *Mathematical Problems in Engineering, 2022,* 4545146. https://doi.org/10.1155/2022/4545146

Cassell, M. K., & Deutsch, A. M. (2023). Urban challenges and the gig economy: How German cities cope with the rise of Airbnb. *German Politics, 32*(2), 319–340. https://doi.org/10.1080/09644008.2020.1719072

Chen, G., Cheng, M., Edwards, D., & Xu, L. (2022). COVİD 19 pandemic exposes the vulnerability of the sharing economy: A novel accounting framework. *Journal of Sustainable Tourism, 30*(5), 1141–1158. https://doi.org/10.1080/09669582.2020.1868484

Clarivate. (2023). *Web of science* [Computer software]. www.webofscience.com/wos

Gursoy, D., Malodia, S., & Dhir, A. (2022). The metaverse in the hospitality and tourism industry: An overview of current trends and future research directions. *Journal of Hospitality Marketing & Management, 31*(5), 527–534. https://doi.org/10.1080/19368623.2022.2072504

Gürsoy, S. (2023). Multidimensional scientometric analysis for the gig economy. *Cumhuriyet Üniversitesi İktisadi ve İdari Bilimler Dergisi, 24*(2), 195–210. https://doi.org/10.37880/cumuiibf.1198210

Hall, C. M., Prayag, G., Safonov, A., Coles, T., Gossling, S., & Koupaei, S. N. (2022). Airbnb and the sharing economy introduction. *Current Issues in Tourism, 25*(19), 3057–3067. https://doi.org/10.1080/13683500.2022.2122418

Jo, H. (2023). Tourism in the digital frontier: A study on user continuance intention in the metaverse. *Information Technology & Tourism, 25*(3), 1–24. https://doi.org/10.1007/s40558-023-00257-w

Kim, N., Park, J., & Choi, J.-J. (2017). Perceptual differences in core competencies between tourism industry practitioners and students using Analytic Hierarchy Process (AHP). *Journal of Hospitality Leisure Sport & Tourism Education, 20,* 76–86. https://doi.org/10.1016/j.jhlste.2017.04.003

Klarin, A., & Suseno, Y. (2021). A state-of-the-art review of the sharing economy: Scientometric mapping of the scholarship. *Journal of Business Research, 126,* 250–262. https://doi.org/10.1016/j.jbusres.2020.12.063

Kobis, N. C., Soraperra, I., & Shalvi, S. (2021). The consequences of participating in the sharing economy: A transparency-based sharing framework. *Journal of Management, 47*(1), 317–343. https://doi.org/10.1177/0149206320967740

Koo, C., Kwon, J., Chung, N., & Kim, J. (2022). Metaverse tourism: Conceptual framework and research propositions. *Current Issues in Tourism, 26*(20), 3268–3274. https://doi.org/10.1080/13683500.2022.2122781

Laurenti, R., Singh, J., Cotrim, J. M., Toni, M., & Sinha, R. (2019). Characterizing the sharing economy state of the research: A systematic map. *Sustainability, 11*(20), 5729. https://doi.org/10.3390/su11205729

Lin, K. J., Ye, H., & Law, R. (2023). Understanding the development of blockchain-empowered metaverse tourism: An institutional perspective. *Information Technology & Tourism*, *25*, 585–603. https://doi.org/10.1007/s40558-023-00262-z

Liu-Lastres, B., Wen, H., & Huang, W.-J. (2023). A reflection on the Great Resignation in the hospitality and tourism industry. *International Journal of Contemporary Hospitality Management*, *35*(1), 235–249. https://doi.org/10.1108/IJCHM-05-2022-0551

Monaco, S., & Sacchi, G. (2023). Travelling the metaverse: Potential benefits and main challenges for tourism sectors and research applications. *Sustainability*, *15*(4), 3348. https://doi.org/10.3390/su15043348

Nemer, D., Spangler, I., & Dye, M. (2018). Airbnb and the costs of emotional labor in Havana, Cuba. *Companion of the 2018 ACM Conference on Computer Supported Cooperative Work and Social Computing (CSCW'18)*, 245–248. https://doi.org/10.1145/3272973.3274066

Nga, J. L. H., Ramlan, W. K., & Naim, S. (2021). Covid 19 pandemic and unemployment in Malaysia: A case study from Sabah. *Cosmopolitan Civil Societies-An Interdisciplinary Journal*, *13*(2), 73–90. https://doi.org/10.5130/ccs.v13.i2.7591

Öz, S. (2023). GIG Ekonomisi ve Emek Piyasaları Bibliometrik Analizi (GIG economy and labour markets: A bibliometric analysis). In *Türk Emek Piyasasında Güncel Tartışmala0r (Murat Kalkan)* (p. 343). Filiz Yayınları.

Saaty, T. L. (1966). A discrete search problem in pattern recognition. *IEEE Transactions on Information Theory*, *12*(1), 69–70. Scopus. https://doi.org/10.1109/TIT.1966.1053850

Saaty, T. L. (1974). Measuring the fuzziness of sets. *Journal of Cybernetics*, *4*(4), 53–61. Scopus. https://doi.org/10.1080/01969727408546075

Sanchez-Amboage, E., Enrique Membiela-Pollan, M., Martinez-Fernandez, V.-A., & Molinillo, S. (2023). Tourism marketing in a Metaverse context: The new reality of European museums on meta. *Museum Management and Curatorship*, *38*(4), 468–489. https://doi.org/10.1080/09647775.2023.2209841

Shaygani, F., Marzaleh, M. A., & Peyravi, M. (2022). Metaverse: A modern approach to medical tourism industry. *Iranian Journal of Public Health*, *51*(12), 2844–2845.

Suanpang, P., Niamsorn, C., Pothipassa, P., Chunhapataragul, T., Netwong, T., & Jermsittiparsert, K. (2022). Extensible metaverse implication for a smart tourism city. *Sustainability*, *14*(21), 14027. https://doi.org/10.3390/su142114027

Tian, C., Peng, J., Zhang, W., Zhang, S., & Wang, J. (2020). Tourism environmental impact assessment based on improved AHP and picture fuzzy PROMETHEE II methods. *Technological and Economic Development of Economy*, *26*(2), 355–378. https://doi.org/10.3846/tede.2019.11413

Tsai, S. (2022). Investigating Metaverse marketing for travel and tourism. *Journal of Vacation Marketing*, *0*(0). https://doi.org/10.1177/13567667221145715

Ucgun, G. O., & Sahin, S. Z. (2023). How does Metaverse affect the tourism industry? Current practices and future forecasts. *Current Issues in Tourism*, *12*(2), 63–78. https://doi.org/10.1080/13683500.2023.2238111

Ulkhaq, M. M., Akshinta, P. Y., Nartadhi, R. L., & Nugroho, S. W. P. (2016). Assessing sustainable rural community tourism using the AHP and TOPSIS approaches under fuzzy environment. In A. J. Arumugham, M. M. Ulkhaq, M. Kocisko, R. K. Goyal, W. A. Yusmawiza, & X. Qiu (Eds.), *2016 3rd international conference on industrial engineering and applications (ICIEA 2016)* (Vol. 68, p. 09003). EDP Sciences. https://doi.org/10.1051/matecconf/20166809003

Volchek, K., & Brysch, A. (2023). Metaverse and tourism: From a new niche to a transformation. In B. Ferrer-Rosell, D. Massimo, & K. Berezina (Eds.), *Information and communication technologies in tourism 2023, Enter 2023* (pp. 300–311). Springer International Publishing AG. https://doi.org/10.1007/978-3-031-25752-0_32

Vosviewer. (2023). *Centre for science and technology studies, Leiden University* (1.6.18) [English]. Centre for Science and Technology Studies, Leiden University. https://vosviewer.com

Wang, X., Li, X., Zhen, F., & Zhang, J. (2016). How smart is your tourist attraction? Measuring tourist preferences of smart tourism attractions via a FCEM-AHP and IPA approach. *Tourism Management, 54*, 309–320. https://doi.org/10.1016/j.tourman.2015.12.003

Wei, W. (2023). A buzzword, a phase or the next chapter for the internet? The status and possibilities of the Metaverse for tourism. *Journal of Hospitality and Tourism Insights, 12*(2), 63–78. https://doi.org/10.1108/JHTI-11-2022-0568

Yang, F. X. X., & Wang, Y. (2023). Rethinking metaverse tourism: A taxonomy and an agenda for future research. *Journal of Hospitality & Tourism Research, 0*(0). https://doi.org/10.1177/10963480231163509

Yang, S., & Kong, X. (2022). Evaluation of rural tourism resources based on AHP-fuzzy mathematical comprehensive model. *Mathematical Problems in Engineering, 2022*, 7196163. https://doi.org/10.1155/2022/7196163

5 Transforming Tourist Consumer Experience Through Virtualization

Does Metaverse Matter?

Alberto Gabriel Ndekwa and Blandina Kisawike

Introduction

One of the service industries with the greatest rate of growth in the world is thought to be tourism, which is also a key driver of development in both established and developing nations. Selimi et al. (2017) noted that since both domestic and foreign visitors require a variety of services during their visits, tourism generates economic opportunities in a number of tourism-related business sectors, including lodging, dining establishments, travel agencies or guides, retail stores, and sports facilities. This suggests that there are excellent prospects for locals to sell goods or provide services to both domestic and foreign tourists. Hafidh and Rashid (2021) promoted the idea that tourism may help create jobs, foreign exchange revenues, and income. In this instance, tourism is widely acknowledged as one of the most noteworthy business prospects with significant social and economic influence. It also constitutes a vital industry for the global economy, particularly in some regions of the world. The results of previous research (Selimi et al., 2017; García & Navarrete, 2022; Ndekwa, 2023) indicate that tourism is becoming more and more important to developing countries and that development organizations are increasingly getting involved in this field.

However, the travel and tourism industries are facing previously unheard-of difficulties as a result of recent global catastrophes and events, such as COVID-19, wars, and climate change (Filho, 2022). Related findings by García and Navarrete (2022) indicate that the hospitality sector is negatively impacted on an economic and social level by factors such as COVID-19, war, and climate change. Al-Mughairi et al. (2021) have identified several economic implications of the situation, such as decreased demand from customers on a national and worldwide scale, interruptions in logistics and distribution networks, and damaged relationships with suppliers, customers, and staff. Narzullayeva and Mukhtarov (2021) have observed a marked decline in both visitor numbers and the export of tourism-related services. These global disasters have caused travel plans to be disrupted, tourism to be shut down, and tourist-related firms to suffer large losses. The majority of tourist-related firms in the case study under examination suffered greatly from the pandemic as a result of national lockdowns and the suspension or cancellation of activities linked to tourism and hospitality (Huynh et al., 2021). These world calamities have led to

DOI: 10.4324/9781003497004-6

tourism shutdowns, disrupted travel plans, and significant losses for businesses in the tourism sector. Findings of other scholars such as Huynh et al. (2021) indicated that the majority of tourism businesses in the examined case study seriously suffered from the pandemic due to national lockdowns, as well as the suspension or cancellation of tourism- and hospitality-related services.

These tendencies had lowered the consumption of tourism products by tourists and even their travel and opportunities to enjoy tourism destinations (Ndekwa, 2021a, 2021b). A number of initiatives were taken to restore tourism consumption and travel. This includes introduction of vaccines and new treatments to travelers to uplift tourism products demand, conflict resorting, and discussions on the impact of climate change (Gvaramadze, 2022). Furthermore, it is well evidenced that some information technologies platforms were adopted before the rise of COVID-19 and other calamities to promote the tourist experience in a virtual and physical manner (Ndekwa, 2014, 2015, 2017; Kavenuke et al., 2017). However, these strategies and technologies did not help a quicken recovery of tourism travel and promotion of tourist consumption and experience due to the fact that their operation was mainly dependent on the physical environment which is considered risky without blended physical and virtual experience (Alvarez-Risco et al., 2022; Weking et al., 2023). New hopes for quick recovery came after the deployment of metaverse in tourism which enables tourists to get experiences in both virtual and physical environments regardless of the distance covered (Pestek & Sarvan, 2020; Alyahya & McLean, 2021).

In the metaverse-based tourism community, Buhalis et al. (2023) noted that the metaverse provides the potential for time travel, enabling users to digitally experience ancient encounters, space explorations, or frightening natural phenomena, including volcano explosions. They further added that users can explore immersive environments for working, learning, transacting, exploring interests, and socializing with others in a virtual context. On the other hand, several authors have pointed out that tourism products and services are perfectly suitable for virtual reality as they have a higher level of intangibility, involvement, and differentiation than other tangible consumer goods, being therefore, more easily sold on the web (Alyahya & McLean, 2021; Aharon et al., 2022). According to Lo and Cheng (2020), metaverse technology's interactive, visual, and immersive features can provide users with virtual experiences in simulated tourist locations to help them plan and make travel-related decisions. Hence, virtualization via metaverse enhances tourist consumer experiences.

In spite of the promises documented justifying how virtualization via metaverse could help in recovery of tourism experience and consumption. Yet there is a lot of fragmented evidence about the linkage of metaverse and tourist experience. Notably, for example, Tsai (2022) found and concluded that metaverse is likely to be accepted by tech-ready and tech-competent users and not merely tourists who are not professional users. Their argument is based on the fact that metaverse is still new and is a concept in tourism and cannot guarantee the successful deployment of tourist virtualization. Other scholars such as Godovykh and Tasci (2020) have argued that the pathway for handling tourist experience has not been clearly set forth due to divergent conceptualizations and insufficient measures of consumer experience in the metaverse age.

In support of the fragmented idea of the promise offered by metaverse to tourists, Ndekwa and Katunzi (2016) advocated that the Internet has transformed a number of economic sectors, but technology is advancing more quickly than the travel and tourist sector can keep up. This is consistent with the observation made by Oncioiu and Priescu (2022), who observed that online tourism is hampered by a large number of virtual technologies with different properties and functions, as well as the lack of criteria for differentiating virtual technologies.

A number of scholars are advocating how virtual experience is becoming an important component of visitors' experience, while others are observing fragmented ideas. It is surprising to see that less attention has been paid to the influence and effects that these technologies have on the tourist consumer experience, and a corresponding measurement framework for the tourist experience is still lacking. Customer experience and information needs have become a priority for the tourism business in the virtual decade which is supported by Konstantinova (2021), who argued that the success of digitization depends on the capacity of the tourism sector to share, learn, and collaborate through a virtual environment. This study filled the gap by analyzing how the transformation of tourist consumer experience through virtualization has brought recovery indication in the metaverse age.

Review of Relevant Literature

Literary Origin of the Term Metaverse

The word "Metaverse" was coined over 30 years ago by renowned science fiction writer Neal Stephenson in his book Snow Crash in 1992. He created the term "metaverse" to refer to a technology that he saw fusing social media, augmented reality, and virtual reality. Investors are finding the metaverse to be an intriguing large-scale technology these days. Barrera and Shah (2023) define the term "metaverse" as the fusion of real-world elements with virtual spaces. This means that rather than being a strictly virtual world, the metaverse is better characterized as an environment that falls along the continuum of extended reality. In order to build a digital world that can be viewed and interacted with in real time, this technology makes use of both virtual reality and augmented reality.

In the metaverse, users can feel life in meta by "entering" the ecosystem. Therefore, it can be said that the metaverse is an interconnected virtual community with no end. The primary objective of the metaverse is to make users interact virtually, overcome the limitations of devices, and immerse themselves in a new world where the boundaries between what is physical and what is digital are increasingly blurred. Lee et al. (2021) pointed out that metaverse is a virtual world that combines physical and digital elements which is made possible by the convergence of Internet and Web technologies. Furthermore, the metaverse, according to Weking et al. (2023), is multiuser environment merging physical reality with digital virtuality. The idea of metaverse is well presented in Figure 5.1.

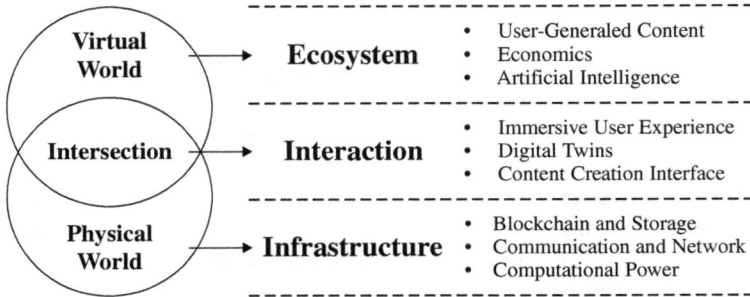

Figure 5.1 Three layers building the metaverse.

According to academics (Lee et al., 2021; Huang et al., 2023), the metaverse is a synthesis of multiple immersive technologies, including mixed reality (MR), AR, VR, and blockchain, rather than a single platform or technology. By enabling real-time interaction between users and digital surroundings, these technologies are making it harder to distinguish between the actual and virtual worlds. Users can create avatars, interact with others, play games, go to events, and take part in a variety of activities in the metaverse, which makes for an immersive and rich experience. The metaverse landscape is being shaped in large part by AI, blockchain, AR, and VR. This indicates that while the fundamental technologies that make up metaverses may be characterized, the idea of the metaverse is still developing, with a variety of technologies playing a role. Although many academics have proposed a number of key technologies, there is not one that is widely accepted. According to Nagarajan et al. (2023), the following important technologies are frequently connected to the metaverse:

Virtual Reality (VR): Users can fully submerge themselves in a computer-generated virtual environment through the use of VR. Usually, it entails donning a headset that monitors head movements and projects audio and graphics to provide the impression of being there in a virtual environment.
Augmented Reality (AR): AR improves the user's sense of reality by superimposing digital information on the physical world. AR can be accessed through wearable technology such as smart glasses or mobile devices.
Mixed reality (MR): MR allows digital material to interact with and coexist with the real environment by fusing aspects of virtual reality and augmented reality. While interacting with virtual things, users can maintain awareness of their physical environment.
Blockchain Technology: A key component of the metaverse, blockchain is renowned for being decentralized and unchangeable.
Blockchain Technology: Known for its immutable and decentralized structure, blockchain technology is essential to the metaverse because it makes virtual

asset ownership, management, and trading safe. It can verify provenance, inter-operability, and digital scarcity in virtual environments.

Artificial Intelligence (AI): AI technologies facilitate natural language interactions, help create intelligent and realistic virtual entities, and improve user experiences in the metaverse as a whole. Chatbots, intelligent automation systems, NPCs (non-player characters), and lifelike avatars can all be powered by AI.

Cloud Computing: A significant amount of processing and storage capacity is needed for the metaverse. The scalability and infrastructure required to manage the enormous volumes of data and intricate calculations required for the immersive experiences offered by the metaverse are made possible by cloud computing.

3D Modeling and Simulation: Tools for 3D modeling and simulation are needed to create virtual environments and metaverse items. The creation, development, and visualization of virtual worlds, characters, and objects are made possible by these technologies.

To understand metaverse marketing, one must know the metaverse and who the audience is. It is an extended reality where physical and digital worlds merge to influence the way people shop, socialize, learn, play, work, and communicate with each other.

Virtual Tourism and Technology Behind

An inventive method of exploring places and experiencing travel without physically being there is virtual tourism. Virtual tourism, as defined by Siddiqui et al. (2022), is the use of digital technologies, such as AR and VR, to explore and experience various countries and cultures without physically traveling there. These technologies have been applied in a range of sectors, such as tourism, entertainment, education, healthcare, and retail, according to Rancati and d'Agata (2022). This is because these technologies have demonstrated efficacy in augmenting learning experiences, boosting productivity, and offering distinctive marketing opportunities. According to Murti et al. (2023), these technologies might be utilized to build virtual versions of locations, giving visitors the opportunity to experience a place without really going there and, thus, lowering the number of tourists that visit those locations. On the other hand, these applications allow users to explore tourist attractions, attend concerts, and do historical tours only via smartphones and personal.

Using this application, tourists can carry out various interactions, such as sending short messages, visiting other tourists' homes, attending communal events, and purchasing as well. They can also visit virtual representations of popular tourist destinations, including museums and landmarks, for instance, a virtual replica of the Louvre Museum. Tourists can virtually explore and see famous works of art at the museum and various virtual locations, such as cities, beaches, and outer space, in real time with other tourists without physically visiting. Hence, it is proven to increase tourist engagement and motivation to explore destinations and experiences virtually.

Relevant Theory

To study how one can experience a phenomenal, social learning theory was proposed in the 1960s and, in 1986, which states that learning in people occurs in a social context as a reciprocal and dynamic interaction of the person, behavior, and environment (Bandura, 1986). This means that experience is gained by learning through social media metaverse of the tourist consumer and environment. The theory aims to explain how people through virtualization regulate their behavior toward experiencing what they interact in the tourist environment via metaverse.

Social learning theory emphasizes the reciprocal relationship between social characteristics of the environment, how they are perceived by individuals, and how motivated and able a person is to reproduce behaviors they see happening around them (Driscoll, 1994). By creating virtual destinations that mimic real-world locations, tourists can experience and learn the culture and beauty of a place without physically visiting it.

Social learning theory is increasingly cited (Gursoy et al., 2022; Go & Kang, 2023) as an essential tool to explain how experience is learned through interacting with the social environment in this case the metaverse. Notably, Pellegrino et al. (2023) used the idea of social learning theory to explain how metaverse influences experience and learning. They found and elaborated how interactive experiences can be designed via metaverse that requires customers to actively participate in activities, such as playing games or completing puzzles. This can help create a more engaging and entertaining experience. Interactive experiences also allow tourists to be more actively involved in their travels, as opposed to simply observing from afar.

In this study, social learning theory was used to analyze how tourist consumer experience is cultivated via learning through virtualization in a metaverse environment. For that reason, the social element of social learning theory has been included where tourist consumers can learn new destinations and behaviors by watching some essential metaverse tools attached to the tourism environment. Hence, this study used this theory to analyze how experience is gained by tourists through the use of metaverse for virtualization.

Relevant Empirical Literature Review

Simoni et al. (2022) in their study of analyzing the influence of virtual reality on promotion of tourist consumer experiance. Findings have revealed the contribution of virtual technologies in promoting the tourist consumer experience through its interactivity. They added that through metaverse, tourists' consumers can undergo pre-travel virtual experience by watching videos prepared with virtual reality before they travel.

In an unrelated perspective, Pellegrino et al. (2023) demonstrated virtualization via metaverse provides more accurate information about interesting places to visit. They added that metaverse enables tourist consumers fairness information and enhances consumer decisions before their travel.

In their study of metaverse in tourism, Buhalis et al. (2023) found and concluded that the introduction of metaverse offers plenty of excitement and a great opportunity for innovation and imagination for both tourism consumers and organizations. Blending the physical and virtual worlds and enabling travelers to operate on both, seamlessly, create great opportunities and challenges for the tourism industry.

On the other hand, Godovykh and Tasci (2020) found that virtualization is a new way to profile tourist consumer interaction with tourist products in a destination setting. While at the destination, tourists can enrich their experiences by receiving personalized services prepared with AI. After the trip, tourists can create electronic word-of-mouth communication by sharing their experiences in interactive social media applications.

Furthermore, Rancati and d'Agata (2022) found and concluded digital metaverse has enabled tourists to become more demanding and independent as they are able to communicate directly via metaverse with basic service providers, bypassing traditional intermediaries and using new online travel intermediaries. They added that metaverse provides opportunities to tourists by improving their customer journey and the path that each tourist consumer takes before making a purchase.

The findings of Sánchez-Amboage et al. (2023) have revealed the significant influence of hybrid communication (physical and virtual) on museums and acted as a pilot experience for their activities. They further found that metaverse contexts tend to reconfigure and strengthen their digital communication and marketing strategies to connect with their audience by virtual means only.

Adel (2023) carried out a study on the significant influence of metaverse tourist interactive experience. The findings from quantitative data revealed that metaverse has a significant influence on the transformation of the tourism sector by enhancing the tourist consumer experience and creating lasting memories for visitors.

Monaco and Sacchi (2023) intend to advance knowledge on how simulated environments could be used to enhance the tourist experience through metaverse. The study results revealed metaverse has a significant contribution to the creation of shared value, especially in situations where tourists are asked to express their expectations and preferences before and after their travel.

Kouroupi and Metaxas (2023) noted that metaverse facilitates tourists' experiences of destinations without physically visiting them to the destination facilities. They added that virtual experiences can offer accessibility to individuals with physical disabilities or financial limitations who may not have been able to travel otherwise. They recommended that virtual experiences be used in a complementary way to enhance the overall tourism experience. However, this study was carried out in Western culture, which calls for a study to be conducted in sub-Saharan African countries to provide further evidence of the influence of metaverse on the tourist experience.

Virtualization Metaverse	Tourist Consumer Experience
VM1: Information sharing	TCE1: Virtual tour
VM2: Digital show case	TCE2: Virtual guide and information
VM3: Simulation	TCE3: Virtual transaction
VM4: Digital information center	TCE4: Customization
VM5: Digital virtual community	TCE5: Blended virtual and physical word
VM6: Digital physical community	
VM7: Digital interaction	

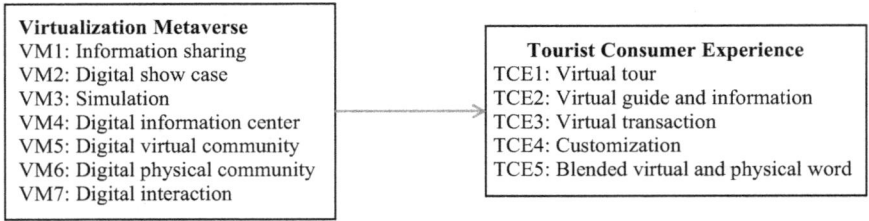

Figure 5.2 Proposed theoretical framework.

Proposed Theoretical Framework

The research framework considers social learning as a leading theory in informing the researcher about the prediction of virtualization via metaverse on tourist consumers. The research framework details the significant influence of virtualization via metaverse on the tourist consumer experience as presented in Figure 5.2.

Methodology

Research Paradigm, Approach, and Design

Research Paradigm is a set of assumptions one must follow in the process of conducting research in order to be able to understand the reality, acquire knowledge, and apply relevant methods. In this study, a positivism paradigm set of assumptions was deployed. According to Creswell (2014), positivism paradigm argues that the reality of the phenomenal and knowledge acquisition about the truth of the phenomenal are originated in the premise of theories and prior empirical literature review. Given these assumptions, in this study, the development of hypothesis and questionnaire was based on the use of relevant theories available and previous empirical studies. This means that the reality and knowledge acquisition of the influence of virtualization through metaverse on tourist consumer experience can be observed from theories and prior empirical studies. The theories and empirical studies enabled the researchers to draw hypothesis and capture data relevant to the study objective. On the other hand, positivism is a paradigm that suggests that knowledge and reality lay on the available theories and empirical evidence and can be studied in quantifiable observation (Saunders et al., 2012). This assumption in the current study was applicable as it allows researchers to use methods that can apply quantitative methods in the process of drawing inference.

In this sense of quantification, this study applied a quantitative research approach in order to be able to analyze statistical data and test hypothesis on the influence of virtualization through metaverse on tourist consumer experience. Scholars such as Apuke (2017) have pointed out that the quantitative research approach is applicable when a researcher is applying the positivism paradigm to establish a significant relationship between the influence of one variable on the other variables.

The research design applied in this study was explanatory. According to Saunders et al. (2012), explanatory research design tends to offer a chance for a researcher to be able to test for cause-and-effect relationships among variables. This study was designed to determine how virtualization via metaverse can cause and effect the tourist consumer experience. In this regard, explanatory served the purpose of this study in analyzing the cause-and-effect relationship between virtualization through metaverse and tourist consumer experience.

Population, Sampling, and Simple Size

This study was designed to analyze tourist's consumer experience by using metaverse in virtualization. Based on this objective, the survey population was tourist consumers who have experience of using metaverse in the tourism sector. The suitability of using tourist consumers as the study population is based on the fact that they are the ones who experience the use of metaverse in developing their tour experience. As argued by Creswell (2014), a population must have common characteristics that are mutually inclusive for all members of the population. In our case, tourists have common characteristics with similar information to provide evidence on the significant influence of metaverse on the tourist experience.

Based on this population, this study used simple random sampling in the sampling of respondents during data collection. The suitability of simple random sampling in this study is that it allows every tourist to have an equal chance of being included during data collection. This is due to the fact that all tourists possess the same experience of using virtualization through metaverse in the context of Tanzania. Saunders et al. (2012) argue that when the population is similar without some cluster, then simple random sampling can be good for picking a sample for the study. On the other hand, simple random sampling is simple to apply and can be used and integrated with other complex sampling techniques.

To get the sample size, this study used the infinity population formula to calculate the sample. This is due to the fact that it is difficult to have the actual number of tourists for the determination of sample size in the context of Tanzania. According to that if the population is not known or is infinity, the researcher can use the sample size formula for an infinity population to calculate the sample. In our case, it is difficult to obtain the actual number of tourists; hence, the formula of simple size for infinite population served the purpose.

Survey Area and Procedure

The study was conducted in Tanzania. The selection of Tanzania is due to the availability of many tourisms' attraction in which a number of tourists are visiting including national parks, heritage sites, and museums. The tourists were selected randomly from tourist entries such as national park, heritage, and museum for the purpose of filling out the questionnaire. The respondents of this study were given brief information about the topic before they fill the questionnaire. They were then asked to filled the questionnaire which took about 10 minutes of their time to fill.

Instrument Design

In this study, the target was to test the influence of virtualization through metaverse on promoting tourist consumer experience. The nature of this study is positivist in nature which requires statistical and measurable data. Given this context, a standardized questionnaire was selected to serve the purpose of capturing standardized data for hypothesis analysis. As argued by Roopa and Rani (2012), a questionnaire is a standardized tool that is good for standardized data in a quantifiable format. The questionnaire was designed with two parts. One was a demographic part and the second consisted of two major themes, namely, virtualization through metaverse as the dependent variable and tourist consumer experience as the independent variable. These are described in Table 5.1 and were measured on a 5-point Likert scale, where 1 represents strongly disagree and 5 represents strongly agree.

The items in Table 5.1 were first developed and written in English, and afterward, translated into Swahili, back-translated, and then retranslated to ensure comparability of data between the English and Swahili versions of the questionnaire. A total of 287 fully completed questionnaires were collected, validated, and included in the analysis as presented in the next section.

Table 5.1 Major constructs and items

Constructs	Items	Authors
Virtualization via metaverse	VM1:Information sharing interesting VM2:Digital showcase VM3:Simulation in both physical and VM4:Digital information center VM5:Digital Virtual community VM6:Digital physical community VM7:Digital interaction	
Tourist consumer experience	TCE1:Virtual tour TCE2:Virtual guide and information TCE3:Virtual transaction TCE4:Customization **TCE4:Blended virtual and physical world**	

Source: Synthesis from literature

Findings and Interpretations of the Results

Model Validation

After data collection, as a first step, we developed a measurement model and performed a PLS-SEM algorithm for the overall model using the Smart PLS 4 software. The first run of the algorism in Smart PLS 4 indicated that a model is not performing well due to the fact that the following measurable variables of independent variables, namely, VM2: Digital showcase and VM5: Digital virtual community, and one measurable variable of dependent variable, namely, TCE5: Blended virtual and physical world had low loading of less than 0.7. The low loading of less than 0.7 affects model performance negatively. The criterion to improve the model performance was argued by Hair et al. (2022) that if measurable variables have scored loading of less than 0.7 is an element for deletion from a model. Using the same criteria, the following variable was removed from the analysis to enhance model performance. After the removal of the mentioned measurable variable then the second run of the PLS-SEM algorithm confirmed good model performance as described in Figure 5.3.

Further analysis of the score produced from the reflective measurement model was done using the following output which was produced during the run of the measurement model.

Table 5.2 shows that each measurable variable has a loading of 0.7 and above. This means that all measurable variables have produced the recommended value for a model to perform better as recommended by Hair et al. (2016).

Further analysis of the quality of the measurement model was done to assess the reliability and convergence validity of the study model. The reliability of the scales was evaluated using Cronbach's alpha and composite reliability, with scores ranging from 0.70 to 0.95 for both Cronbach's alpha and composite reliability, suggesting a good internal consistency of the scales (Liengaard et al., 2021), whereas convergent validity of the constructs was evaluated using the average variance extracted (AVE) criterion, whose values exceeded the minimum limit of 0.50 (Hair

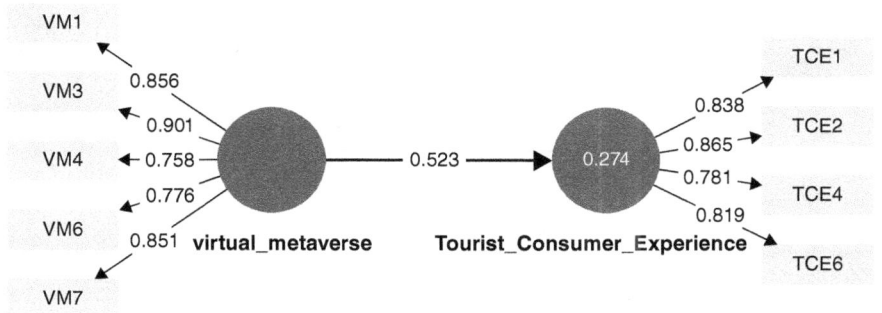

Figure 5.3 Reflective measurement model.

et al., 2022). Both composite reliability and construct AVE values serve as a basis to ensure that the constructs of the proposed model demonstrate good quality.

Table 5.3 shows that the score for both Cronbach's alpha and composite reliability indicates good internal consistency of the model by producing reliable results ranging from 0.848 to 0.917 for both Cronbach's alpha and composite reliability. On the other hand, the results have produced AVE ranging from 0.683 to 0.689 for all constructs of the model, which implies that the measurable variables of each construct have converged well.

We then ran the PLS algorithm to assess discriminant validity through Fornell and Larcker's criterion, in which the square root of the AVE (highlighted in bold on the diagonal in Table 5.4) must be greater than the correlations between constructs in the model. Table 5.4 shows the square root of the AVE of the bold on the diagonal has a value greater than the correlations, demonstrating strong support for discriminant validity acceptance. This is consistent with the suggestion made by Fornell and Larcker (1981), who recommended that the square root of the AVE (on the diagonal) produced in a model must be higher than the correlations observed between constructs of the output (off the diagonal).

Table 5.2 Measurable variable loading

Measurable variables	Tourist_consumer_experience	Virtual_metaverse
TCE1	0.838	
TCE2	0.865	
TCE4	0.781	
TCE6	0.819	
VM1		0.856
VM3		0.901
VM4		0.758

Table 5.3 Reliability and convergence validity

Constructs	Cronbach's alpha	Composite reliability (rho_a)	Composite reliability (rho_c)	AVE
Tourist_con-sumer_experi-ence	0.848	0.864	0.896	0.683
Virtual_metaverse	0.888	0.906	0.917	0.689

Table 5.4 Fornell–Larcker criterion

Constructs	Tourist_consumer_experience	Virtual_metaverse
Tourist_consumer_experience	**0.827**	
Virtual_metaverse	0.523	**0.830**

Table 5.5 HTMT ratio

	HTMT ratio
Virtual_metaverse -> tourist_consumer_experience	0.579

In addition, we used the criterion of the heterotrait–monotrait (HTMT) ratio of correlations, in which the ratio between the constructs is expected to be less than 0.80, which was also met in this research (Hair et al., 2017).

The results for the HTMT ratio in Table 5.5 have produced a value of 0.59 which indicates good discriminant validity. As argued by Liengaard et al. (2021), any value less than 0.8 of the HTMT ratio in a model indicates that the model does not suffer from discriminant validity problems. This implies that measurable variables measuring different constructs do not correlate well but those measuring similar constructs are correlating well.

Empirical Results of Hypothesis Testing

This study used an analysis of variance to test the significant influence of the independent variable of the hypothesis that virtualization through metaverse has influence on tourist consumer experience. In this part of hypothesis testing, the structural model is evaluated using three criteria, namely, VIP, R square, and path coefficient as elaborated as follows.

Starting with VIP criteria, this criterion was used to evaluate if the model is suffering from collinearity where each measurable variable was assessed to see if have scored the VIP value of less than 0.5 (Hair et al., 2016). In Table 5.6, the results have demonstrated that each measurable variable used in the model has scored a VIP of less than 0.5. These results imply that the model is not suffering from collinearity. The results further imply that all problems associated with collinearity are well addressed in the study and the model is not suffering from multi-collinearity.

Further evaluation of the structural model was done to determine exploratory power and predictive power using R square. According to Hair et al. (2022), a model is accepted if the dependent variable has scored an R square of less than 0.8. The output presented in Table 5.7 indicates that the model has good predictive power and explanatory power with an R square of 0.274.

After the assessment of VIP and R square, further analysis was done to determine the model path coefficient using a significant p-value. Scholars have recommended a p-value for the path coefficient to be less than 0.5 or equal for the hypothesis to be accepted. Figure 5.4 shows that the p-value of the independent variable (virtualization via metaverse) on tourist consumer experience is 0.00. This value indicates a significant influence of the independent variable, namely, virtualization through metaverse on tourist consumer experience based on the recommendation made by Hair et al. (2016), that any hypothetical relationship with a significant value less

Table 5.6 Results for evaluation of collinearity

Measurable variable	VIF
TCE1	2.338
TCE2	2.472
TCE4	3.162
TCE6	3.392
VM1	2.775
VM3	3.762
VM4	3.203
VM6	3.312
VM7	2.639

Table 5.7 Explanatory and predictive power

Independent variable (construct)	R-square	R-square adjusted
Tourist_consumer_experience	0.274	0.272

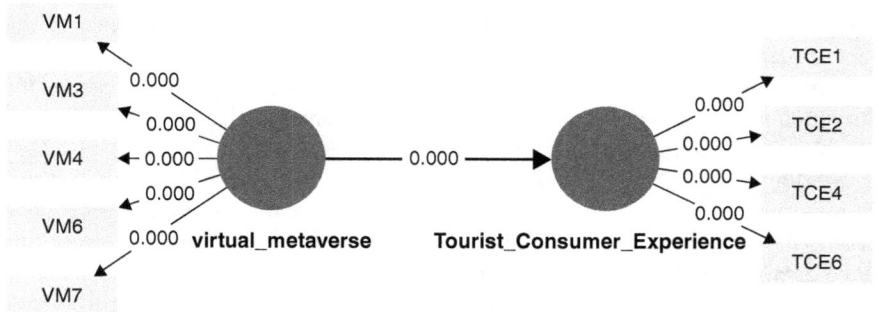

Figure 5.4 Structural model.

than 0.05 or equal is considered significant. From Figure 5.4, it is indicated that all measurable variables used in the model to measure the independent valuable virtualization have produced a significant p-value less than 0.05.

Interpretation of the Results

In the test hypothesis, the value of the prediction of the independent variable, namely, virtualization via metaverse to the dependent variable, namely, tourist consumer experience (p < 0.00), is less than 0.05 and confirms that the virtualization via metaverse has a significant impact on the tourist consumer experience. This further confirms that the practice of metaverse technology in tourism can facilitate the tourist consumer experience. In addition, virtualization is gradually increasing the number of tourists who demonstrate that virtualization via metaverse can play the strongest role in tourism for demonstrating tourist consumer experience.

Discussion

In this study, it was hypothesized that virtualization through metaverse has a significant influence on the tourist consumer experience. The findings of this study have indicated the significant influence of virtualization via metaverse on the tourist consumer experience. The findings have explained the influence of metaverse is due to its explanatory power that it offers virtual tour, virtual guide and information, customization, and blended virtual and physical worlds. These findings concur with the argument made by Oncioiu and Priescu (2022) that there is great potential for the use of virtualization metaverse in the tourism sector which allows tourists to communicate, interact, or explore the virtual world. The findings have collaborated and supported social learning theory which states that learning occurs in a social environment as a dynamic interaction of the person and environment. This means that tourist consumers learn to get experience of destinations in a metaverse environment as a reciprocal and dynamic interaction of the tourist consumers and destination environment. This implies that theories that were validated using Western data can also be applied in sub-Saharan African countries to study the linkage between metaverse and consumer experience.

This study has aligned with the empirical observation by Monaco and Sacchi (2023), who observed that metaverse enables tourists to experience the physical world via the virtual world. The findings also support the argument made by Israfilzade (2022) that metaverse provides an opportunity to build virtual experiences that connect with physical world activities. They added that through metaverse many tourist services that are not possible to be accessed in the physical world can be accessed. These findings imply that by deploying in a virtual destination site, tourists can be able to explore their destination of choice.

This study is supported by prior studies, but they differ in explaining how virtualization via metaverse impacts the tourist experience. In this study, the explanation why virtualization via metaverse contributes to the tourist experience is explained by the ability of metaverse to offer information sharing, digital information center, simulation of both physical and virtual worlds, digital physical community, and digital interaction. In other studies, for example, Hananto et al. (2024) found that tourist consumer experiences in metaverse are explained by customization and co-creation of blended virtual and physical experiences, which in turn allows consumers to interact with other tourist consumers before, during, and after their visit, while Basori et al. (2023) explained that virtualization through metaverse stimulates the package offered at hotels. They added that tourist consumers are able to take virtual tours which gave them an experience of staying at the hotels and visiting the cities and the surrounding areas. Their findings imply that tourists will be able to interact with the travel agency, and when they want to book a room at the hotel, they will be able to see employees preparing the room, walk around the hotel right at the time of booking, and much more via metaverse technology. Hence, metaverse is a powerful tool for building experience between the tourist and the brand.

Contrary to the findings of this study, some studies, for example, Fazio et al. (2023), found insignificant contribution of metaverse on tourist consumer experiences. Fazio et al. (2023) argued that the insignificant contribution is due to the fact that some tourist consumers are not aware and do not possess adequate knowledge of the metaverse simply because the metaverse is still in a moment of early diffusion in the tourism sector.

Conclusion and Implication

These study findings have proved that virtualization through metaverse has a significant contribution to promoting the tourist experience in any destination. It is therefore concluded that virtualization through metaverse promotes the tourist experience in any destination. This is explained by the fact that metaverse has an important option of allowing a tourist to conduct a virtual tour before the time of traveling. Metaverse tends to provide a room for tourists to have a social learning of tourism products and services in any destination that he/she is planning to travel to. While in the problem formulation of this study, it was found that metaverse offers a fragmented explanation, the findings of this study imply that the fragmented explanation of the concept of metaverse is due to its flexibility which is an advantage that metaverse can be applied to offer virtual services in any tourism department and its application can be captured in all phases of tourism value chain.

References

Adel, M. A. (2023). The role of metaverse to create an interactive experience for tourists. *Journal of Association of Arab Universities*, 24(1), 242–269.

Aharon, D. Y., Demir, E., & Siev, S. (2022). Real returns from unreal world? Market reaction to *Metaverse* disclosures. *Research in International Business and Finance*, *63*, 101778. https://doi.org/10.1016/j.ribaf.2022.101778

Al-Mughairi, H. M. S., Bhaskar, P., & Alazri, A. K. H. (2021). The economic and social impact of COVID-19 on tourism and hospitality industry: A case study from Oman. *Journal of Public Affairs*, e2786, 1–11. https://doi.org/doi.org/10.1002/pa.2786

Alvarez-Risco, A., Del-Aguila-Arcentales, S., Rosen, M. A., & Yáñez, J. A. (2022). Social cognitive theory to assess the intention to participate in the Facebook metaverse by citizens in Peru during the COVID-19 pandemic. *Journal of Open Innovation, Technology, Marketing and Complex*, *8*, 142. https://doi.org/10.3390/joitmc8030142

Alyahya, M., & McLean, G. (2021). Examining tourism consumers' attitudes and the role of sensory information in virtual reality experiences of a tourist destination. *Journal of Travel Research*, 1–16.

Apuke, O. D. (2017). Quantitative research methods: A synopsis approach. *Arabian Journal of Business and Management Review*, *6*(10). https://doi.org/10.12816/0040336

Bandura, A. (1986). *Social foundations of thought and action*. Prentice Hall.

Barrera, K. G., & Shah, D. (2023). Marketing in the Metaverse: Conceptual understanding, framework, and research agenda. *Journal of Business Research*, *155*, 113420.

Basori, M. H., Triyono, T., Hidayat, M. N., Dewi, U. F., & Sensato, S. T. (2023). Metaverse in tourism as a medium for conservation of maritime area in Karimunjawa National Park. *Jurnal Audiens*, *2*(2), 221–230. https://doi.org/10.18196/jas.v4i2.19

Buhalis, D., Lin, M. S., & Leung, D. (2023). Metaverse as a driver for customer experience and value co-creation: Implications for hospitality and tourism management and marketing. *International Journal of Contemporary Hospitality Management, 35*(2), 701–716. https://doi.org/10.1108/IJCHM-05-2022-0631

Creswell, J. W. (2014). *Research design: Qualitative, quantitative and mixed methods approaches* (4th ed.). Sage.

Driscoll, J. (1994). Reflective practice in practise. *Senior Nurse, 13*(7), 47–50.

Fazio, G., Fricano, S., Lannolino, S., & Pirrone, C. (2023). Metaverse and tourism development: Issues and opportunities in stakeholders' perception. *Information Technology & Tourism, 25*, 507–528. https://doi.org/10.1007/s40558-023-00268-7

Filho, W. L. (2022). Will climate change disrupt the tourism sector? *International Journal of Climate Change Strategies and Management, 14*(2), 212–217. https://doi.org/10.1108/IJCCSM-08-2021-0088

Fornell, C. G., & Larcker, D. F. (1981). Evaluating structural equation models with unobservable variables and measurement error. *Journal of Marketing Research, 18*(1), 39–50.

García, C., & Navarrete, M. C. (2022). Vaccine tourism: A new entrepreneurship. *Journal of Tourism and Heritage Research, 5*(4), 1–10.

Go, H., & Kang, M. (2023). Metaverse tourism for sustainable tourism development: Tourism agenda 2030. *Tourism Review, 86*. https://doi.org/10.1108/TR-02-2022–0102

Godovykh, M., & Tasci, A. D. A. (2020). Customer experience in tourism: A review of definitions, components, and measurements. *Tourism Management Perspectives, 35*, 1–10. https://doi.org/10.1016/j.tmp.2020.100694

Gursoy, D., Malodia, S., & Dhir, A. (2022). The metaverse in the hospitality and tourism industry: An overview of current trends and future research directions. *Journal of Hospitality Marketing & Management, 31*(5), 527–534.

Gvaramadze, A. (2022). Digital technologies and social media in tourism. *European Scientific Journal, 18*(10), 28. https://doi.org/10.19044/esj.2022.v18n10p28

Hafidh, H. A., & Rashid, Z. A. (2021). The impact of tourism development on economic development: The case of Zanzibar. *Asian Journal of Education and Social Studies, 18*(1), 41–50.

Hair, J. F., Babin, B. J., & Krey, N. (2017). Covariance-based structural equation modeling in the journal of advertising: Review and recommendations. *Journal of Advertising, 46*, 163–177. https://doi.org/10.1080/00913367.2017.1281777

Hair, J. F., Hult, G. T. M., Ringle, C. M., & Sarstedt, M. (2022). *A primer on partial least squares structural equation modeling (PLS-SEM)*. Sage.

Hair, J. F., Sarstedt, M., Matthews, L., & Ringle, C. M. (2016). Identifying and treating unobserved heterogeneity with FIMIX-PLS: Part I-method. *European Business Review, 28*(1), 63–76.

Hananto, H., Megawati, V., & Pratono, A. H. (2024). Digital innovation in the tourism industry: Some evidence from Indonesia. *Proceedings of the 20th International Symposium on Management (INSYMA 2023): Series: Advances in Economics, Business and Management Research*, 172–177. https://doi.org/10.2991/978-94-6463-244-6_28

Huang, X-T., Wang, J., Wang, Z., Wang, L., & Cheng, C. (2023). Experimental study on the influence of virtual tourism spatial situation on the tourists' temperature comfort in the context of metaverse. *Frontiers in Psychology, 13*, 1062876. https://doi.org/10.3389/fpsyg.2022.1062876

Huynh, D. V., Truong, T. T. K., Duong, L. H., Nguyen, N. T., Dao, G. V. H., & Canh Ngoc Dao, C. N. (2021). The COVID-19 pandemic and its impacts on tourism business in a developing city: Insight from Vietnam. *Economies, 9*, 172. https://doi.org/10.3390/economies9040172

Israfilzade, K. (2022). Marketing in the metaverse: A sceptical viewpoint of opportunities and future research directions. *The Eurasia Proceedings of Educational & Social Sciences, 24*, 53–60.

Kavenuke, R., Matimbwa, H., Samwel, L., Jummane, H., Kapinga, E., & Ndekwa, A. G. (2017, August 10–11). *Mobile money payment adoption in tourism: Incidence from SMEs from Zanzibar.* 17th International Conference on African Entrepreneurship & Small Business, University of Dar es Salaam Business School, Dar es Salaam Tanzania.

Konstantinova, S. (2021). Digital transformation in tourism. Knowledge. *International Journal, 35*(1), 188–193.

Kouroupi, N., & Metaxas, T. (2023). Can the metaverse and its associated digital tools and technologies provide an opportunity for destinations to address the vulnerability of over-tourism? *Tourism and Hospitality, 4*, 355–373. https://doi.org/10.3390/tourhosp4020022

Lee, L. H., Braud, T., Zhou, P., Wang, L., Xu, D., Lin, Z., & Hui, P. (2021). All one needs to know about metaverse: A complete survey on technological singularity, virtual ecosystem, and research agenda. *arXiv preprint*, arXiv:2110.05352.

Liengaard, B. D., Sharma, P. N., Hult, G. T. M., Jensen, M. B., Sarstedt, M., Hair, J. F., & Ringle, C. M. (2021). Prediction: Coveted, yet forsaken? Introducing a cross-validated predictive ability test in partial least squares path modelling. *Decision Sciences, 52*(2), 362–392

Lo, W. H., & Cheng, K. L. B. (2020). Does virtual reality attract visitors? The mediating effect of presence on consumer response in virtual reality tourism advertising. *Information Technology & Tourism, 22*, 537–562. https://doi.org/10.1007/s40558-020-00190-2

Monaco, S., & Sacchi, G. (2023). Travelling the metaverse: Potential benefits and main challenges for tourism sectors and research applications. *Sustainability, 15*, 3348. https://doi.org/10.3390/su15043348

Murti, K. G. K., Darma, G. S., Mahyuni, L. P., & Gorda, N. E. S. (2023). Immersive experience in the metaverse: Implications for tourism and business. *International Journal of Applied Business Research, 5*(2), 187–207. https://ijabr.polban.ac.id/ijabr/

Nagarajan, G., Moorthy, V. S., Mohamed, A. K., Mohideen, A. S., Ishaq, M. M., & Lakshmi, M. R. (2023). The role of the metaverse in digital marketing. *Journal for Educators, Teachers and Trainers, 14*(5), 51–59.

Narzullayeva, G. S., & Mukhtarov, M. M. (2021). Impact of covid-19 on tourism: The restoration of tourism and the role of young entrepreneurs in IT. *World Economics & Finance Bulletin, 2*, 14–18.

Ndekwa, A. G. (2014). Factors influencing adoption of information and communication technology (ICT) among small and medium scale enterprises (SMEs) in Tanzania. *International Journal of Research in Management and Technology, 4*(5), 273–280.

Ndekwa, A. G. (2015). Determinants of adopter and non adopter of computerizing accounting system (CAS) among small and medium enterprises (SMEs) in Tanzania. *International Journal of Innovative Science, Engineering & Technology, 2*(1), 438–449.

Ndekwa, A. G. (2017). *Factors influencing adoption of mobile money services among small and medium enterprises in Tanzania tourism sector* [PhD thesis, The Open University of Tanzania].

Ndekwa, A. G. (2021a). Connecting mobile marketing events and business performance during covid-19 business crisis in Africa. *International Journal of Marketing & Human Resource Research, 2*(4), 178–188.

Ndekwa, A. G. (2021b). COVID 19 and higher education institution operation: Unpacking the potential of higher education through acceptance of online learning. *International Journal of Education, Teaching, and Social Sciences, 1*(1), 48–58. https://doi.org/10.47747/ijets.v1i1.491

Ndekwa, A. G. (2023). Linking business owner's market capability and mobile marketing adoption: Experience from Tanzania. In M. Valeri (Ed.), *Family business in tourism and hospitality. Tourism, hospitality & event management.* Springer. https://doi.org/10.1007/978-3-031-28053-5_6

Ndekwa, A.G., & Katunzi, T. M. (2016). Small and medium tourist enterprises and social media adoption: Empirical evidence from Tanzanian tourism sector. *International Journal of Business and Management, 11*(4), 71–80.

Oncioiu, I., & Priescu, I. (2022). The use of virtual reality in tourism destinations as a tool to develop tourist behavior perspective. *Sustainability, 14,* 4191. https://doi.org/10.3390/su14074191

Pestek, A., & Sarvan, M. (2021). Virtual reality and modern tourism. *Journal of Tourism Futures, 7*(2), 245–250. https://doi.org/10.1108/JTF-01-2020-0004

Pellegrino, A., Wang, R., & Stasi, A. (2023). Exploring the intersection of sustainable consumption and the Metaverse: A review of current literature and future research directions. *Heliyon, 9*(e19190). https://doi.org/10.1016/j.heliyon.2023.e19190

Rancati, E., & d'Agata, A. (2022). Metaverse in tourism and hospitality: Empirical evidence on generation Z from Italy. *European Scientific Journal, 18*(34), 122–135. https://doi.org/10.19044/esj.2022.v18n34p122

Roopa, S., & Rani, M. S. (2012). Questionnaire designing for a survey. *The Journal of Indian Orthodontic Society, 46*(4), 273–277.

Sánchez-Amboage, E., Membiela-Pollán, M. E., Martínez-Fernández, V., & Molinillo, S. (2023). Tourism marketing in a metaverse context: The new reality of European museums on Meta. *Museum Management and Curatorship, 38*(4), 468–489. https://doi.org/10.108 0/09647775.2023.2209841

Saunders, M., Lewis, P., & Thornhill, A. (2012). *Research methods for business students* (6th ed.). Harlow Pearson Educational Limited.

Selimi, N., Sadiku, L., & Sadiku, M. (2017). The impact of tourism on economic growth in the Western Balkan countries: An empirical analysis. *International Journal of Business and Economic Sciences Applied Research, 10*(2), 19–25. https://doi.org/10.25103/ijbesar.102.02

Siddiqui, M. S., Syed, T. A., Nadeem, A., Nawaz, W., Alkhodre, A. (2022). Virtual tourism and digital heritage: An analysis of VR/AR technologies and applications. *International Journal of Advanced Computer Science and Applications, 13*(7), 305–315.

Simoni, M., Sorrentino, A., Leone, D., & Caporuscio, A. (2022). Boosting the pre-purchase experience through virtual reality. Insights from the cruise industry. *Journal of Hospitality and Tourism Technology, 13*(1), 140–156.

Tsai, S. (2022). Investigating metaverse marketing for travel and tourism. *Journal of Vacation Marketing, 0*(0), 1–10. https://doi.org/10.1177/13567667221145715

Weking, J., Desouza, K. C., Fielt, E., & Kowalkiewicz, M. (2023). Metaverse-enabled entrepreneurship. *Journal of Business Venturing Insights, 19*(2023), e00375. https://doi.org/10.1016/j.jbvi.2023.e00375

6 Metaverse as a Socio-Technical Phenomenon in Hospitality and Tourism

An Approach to Sustainable Development Goals

Seyed Ilia Daneshpour, Nasser Koleini Mamaghani, Hassan Sadeghi Naeini, and Seyed Hashem Mosaddad

Introduction

The metaverse is formed parallel to the real space as a second living space, or as Bobrowsky and Needleman (2022) described, the metaverse, a concept that has already been within the Internet, can bring virtuality beside reality. It has the potential to bring virtual reality to the real world. This is particularly significant for the hospitality and tourism industries, as they strive to create immersive experiences for travelers. However, the liveliness of the metaverse relies on human interaction. Humans possess unique capabilities that shape their experience of the world, and researchers aim to replicate these within the metaverse to match real-world conditions. The advent of Web3 and the metaverse marks a significant technological evolution. It has the power to revolutionize travel experiences and remove barriers for job seekers, regardless of physical abilities or location. The metaverse represents a promising frontier where virtual and tangible realities merge. It has the potential to redefine how we interact with the world and each other. By bridging the gap between physical and virtual realms, the metaverse offers endless possibilities for innovation, growth, and inclusivity. Although the pandemic caused a decrease in the number of travelers, the willingness of tourists to visit popular tourist destinations still put the environment under challenge with all its sustainability concerns (WTTC, 2022). In a prose translation of the *Bustan of Sa'adi* (The famous Persian poet), originally published as part of the *Wisdom of the East series*, it is stated that one was cutting the branches and trunk of a tree (Rajaee, 2000). The future of tourism depends on practices that support the sustainability of tourism resources and communities. Policies and strategies have been formulated to attain this objective (Budeanu et al., 2016). Besides, the travel requests of users in Generation Z, commonly addressed to Zoomers and Alpha generation, as native citizens of the Internet era, will be visible along with the current users and stakeholders soon (Ameen et al., 2022; Scholz & Vyugina, 2019). These users' deep familiarity with related technologies and the spaces developed on Internet platforms, such as virtual and augmented reality, demonstrate the potential of this type of tourism. Gen Z

DOI: 10.4324/9781003497004-7

tourists are influenced by social media marketing when choosing a destination and are more likely to visit a desired location than other generations (Liu et al., 2023). Therefore, it is necessary to address this new market. In this sphere, the tourism and the hospitality industry, the concept of metaverse tourism pertains to immersive environments that represent a fusion of tangible and virtual realms, offering tourists the opportunity to engage in experiential journeys and spatial encounters. A hybrid metaverse-driven tourism that is brought into existence through infrastructural elements and a multitude of sensory stimuli. Creating a comprehensive metaverse tourism ecosystem and research agenda necessitates the development of theoretical frameworks and research propositions. This involves a deep understanding of the advantages derived from integrating the metaverse into the field of tourism (Kwon & Koo, 2022) to provide a satisfying end-user experience. To achieve this goal, various studies and scholarly literature were studied to obtain a selection of terminologies and requirements related to the past and present and the demands for the future. It will be discussed in the following.

Tourism Terminology and Categorizations

Tourists can be classified as domestic or international while also as Individual or group tourists. Besides, tourism can be classified by the aim of traveling as healthcare, visiting, transit, leisure, and professional. Tourism is a crucial industry of national significance, encompassing various activities such as providing transportation, accommodation, recreation, food, and related services for both domestic and international travelers that play a vital role in the economy (McIntosh et al., 1995). One of the first definitions and classifications of tourism was presented by Leiper (1979), who articulated five essential constituents that establish a functional and spatial tourism system: tourists, regions that generate demand, destination regions, a tourism industry, and transit routes.

The terms and concepts that were widely used, such as "e-tourism," "smart tourism," "virtual tourism," and some others, which are not very common or are emerging in literature, such as "mobile tourism," "intelligent tourism," "digital form of tourism," "digital tourism or digitalization of tourism)," "digital free tourism (DFT)," "tourism 4.0," "sustainable tourism," and "circular tourism," which are the definitions related to the research subject, are briefly presented in the following. The integration of information communication technology and latency-less connectivity has made travel and daily life seamless, demonstrating how vital technology has made e-tourism (Shafiee et al., 2019). In other words, the rise of e-commerce has led to the growth of related sectors, including e-tourism, e-marketing, e-finance, and e-production, among others. As Dolnicar et al. (2020) highlighted, e-tourism has been a topic of study since the early 1990s. Over time, it has developed into a significant area of research, focusing on theory and information technology development related to key issues in the tourism realm. As Maha et al. (2012) mentioned, e-tourism thrives on the vital components of central or computerized reservation systems, computerized networks used by travel agents or global distribution systems, hotel or property management systems, and social media.

Smart tourism aims to maximize environmental, cultural, and social values by implementing and applying technologies such as the Internet of Things and leveraging data collected by interconnected sensors with 5G networks. By utilizing these technologies, it is possible to create a more connected and efficient tourism industry that can improve the experience for tourists while also preserving local cultures and environments (I. Egger et al., 2020; J. Li et al., 2018; Y. Li et al., 2017).

As (S. Chen et al., 2022) said, the tourism industry has embraced innovations to enhance the satisfaction of its users while streamlining operational costs, and these are the exact reasons that emphasize the growing importance of understanding technological shifts in travel manners and procedures. This goal is not related and limited to today; Perry Hobson and Williams (1995) noted that the concept of virtual tourism using virtual reality can be traced back to the same time when Stephenson proposed the metaverse. However, according to Cheong (1995), due to the technological limitations and the pioneering nature of this idea, its development was hindered during those early years. Moreover, users' technical requirements and enthusiasm were not sufficient to support its futuristic development. Metaverse and under-developing Web 3.0 can introduce a fresh perspective and paradigm following e-tourism, smart tourism, and virtual tourism.

Because of the daily use of smartphones and mobile devices, the mobile user experience has been improved and led to enhanced overall user satisfaction (O'Sullivan & Grigoras, 2015). Tourists use devices to capture and share their digital experiences, seeking unconventional travel beyond physical activities (Go & Gretzel, 2009).

Digital Transformation

Various studies employ terms such as digital, digitalized, digitalization, and digital transformation. The absence of precise definitions for these terms can result in their interchangeable use. Verhoef et al. (2021) highlight that digitalization pertains to organizations seeking enhancements in their operations by applying information and communication technology (ICT)-based solutions, adding value to streamline their workflow. On the other hand, digitizing is a process that results in converted content from the analog world welcomed to digital format (Culot et al., 2020). Digital transformation represents a profound shift facilitated by the innovative utilization of digital technologies, coupled with the strategic optimization of critical resources and capabilities. Its objective is to enhance and redefine its value proposition for stakeholders in such a way that this transition can be interpreted as a socio-technical phenomenon (Gong & Ribiere, 2021; Hartl & Hess, 2019). This transformation also leads to establishing a fresh business model that impacts all facets of customer engagement and takes advantage of digital tools, technologies, and strategies to boost productivity and maintain competitiveness in the digital era (Rocha et al., 2018).

Not only simulating or reproducing real spaces in the form of a virtual environment but also creating computer-generated spaces as a new possibility can be considered an added value of digital transformation in the field of metaverse-driven tourism. The opportunity for tourists to immerse themselves while visiting protected, forbidden, or even non-existent sites and attractions, which will be possible through VR as a type of metaverse-driven traveling, or probably tourism, is really valuable (I. Egger et al., 2020). Furthermore, VR can remove accessibility barriers for the elderly or disabled travelers (Slater & Sanchez-Vives, 2016), and this point can be considered one of the exact advantages of entering the metaverse into tourism. However, travel experiences on the historical and heritage sites developed based on VR/AR technologies in the format of a virtual environment would not be as authentic as real travel experiences (Mura et al., 2017). Delving into metaverse attributes such as AR, lifelogging, mirror worlds, and VR has not only enriched our understanding of diverse industries but also endowed them with the potential to offer an entirely new dimension of promoted experiences, especially in tourism and traveling (Filieri et al., 2021). Besides, ensuring effective user interaction is as important as keeping up with tech advancements. However, the tourism industry and activists should prioritize staying up-to-date. In the subsequent section, the chapter will delve into the definitions derived from articles and studies pertaining to this novel environment and domain.

Metaverse

The metaverse is the name that has been chosen for the immersive virtual world imagined by Neal Stephenson in his visionary novel 1992, "Snow Crash." Following Stephenson, in 2011, another book written by Ernest Cline named "Ready Player One," a virtual world is depicted as an oasis. This is a complex concept, which can be interpreted as a multi-user virtual world. The metaverse, coined from the fusion of "meta" (signifying beyond) and "universe," refers to a three-dimensional virtual environment that employs the metaphor of our tangible reality (Lee et al., 2021). In the final two decades of the 20th century, akin to the century preceding the Industrial Revolution, a spectrum of concepts and theories emerged to probe the transition from concrete spaces such as neighborhoods, cities, regions, and countries to the digital domain. Stephenson's notion of a transferred human from tangible spaces to digital realms was not unique, as this trend has continued before and after him. These ideas, rooted in the digital sphere, encompass a selection of terms and concepts briefly alluded to. Noteworthy examples include Toffler's Electronic Cottage in 1981, Fishman's Technoburb in 1987, Castells' Space of Flows in Information City proposed in 1989, Baty's Invisible City and Intelligent City in 1990, Fathi's Telecity introduced in 1991, Reality–Virtuality (RV) continuum of Milgram & Kishino in 1994, Ascher's Metapolis concept in 1995, Mitchell's City of Bits proposed in 1996, Atkinson's Information-age Metropolis in 1996, Donath's Virtual City in 1997, Mitchel's E-Topia introduced in 2000, Townsend's Network City in 2001, Castells' Space of Flows and Places in 2004, Graham's coining of Cybercity in 2004, De Souza e Silva's Hybrid Space in 2006, and Meta (formerly Facebook) try to construct its unique Metaverse in 2021.

The metaverse concept has been explored in various studies, and as Duan et al. (2021) said, due to its emerging notion, the definition, architecture, and mechanism are still in the early stages of development. From the beginning of this notion, most scholars' points of view and understanding can be classified into two poles. In the early years of the emergence of this concept, some of them (Perry Hobson & Williams, 1995) thought it was just a way to escape from the real world and transport to the unreal, imaginary environment simulated by the computer. However, others (Desai et al., 2014; Guttentag, 2010) believe that the developed metaverse-driven environment is not only a way to escape real-world limitations and follow their desires but also, besides the scene of physicality, the virtual stimuli that encourage the users to be in a virtual environment can be likened to an added plug-in or add-on to their real-world experiences. The metaverse can be said to comprise four primary and essential dimensions: the intimate dimension versus external and augmentation versus simulation. This combination's output can appear in different forms. AR superimposes digital creation onto the physical-world artifact. Mirror worlds (MW) as the physical world's reflection. Virtual worlds (VW) as a real-world alternative, mixed reality (MR) in the prior research mentioned, which was interpreted as the fusion of VR and AR simultaneously and lifelogging. Farshid et al. (2018), in following Milgram and Kishino (1994), describe an environment containing the real world and virtual reality, in which the mixed reality is positioned at the center of a continuum between AR and augmented virtuality (AV). In other words, they believe that AR and VR are subsets of MR. Milgram and colleagues introduced the reality–virtuality (RV) continuum notion, spanning from physical reality to virtual environments. They highlighted its significance as a framework for comprehending mixed reality, incorporating real-world and virtual elements. This concept also involved crucial terms like Extent of World Knowledge, which aimed to position and define objects from unmodeled to fully modeled worlds in response to "Where, What, and Where+What" queries. Additionally, they introduced Reproduction Fidelity and Extent of Presence Metaphor (Milgram et al., 1995). Some scholars have recently proposed a term called extended reality (XR), which is being compared with an umbrella. This is because XR encompasses the environment of symbiosis and interaction in both the virtual and real worlds, along with all the digital and physical objects within them (Kunkel & Soechtig, 2017). Or, in another definition, their construction involves sophisticated virtual mapping, modeling, annotation tools, geospatial and other sensors, location-aware, and other lifelogging technologies (Choi & Kim, 2017; Gursoy et al., 2022; Lee et al., 2021). The concept of virtual reality metaverses, which is defined by Makransky and Mayer (2022), has three essential characteristics: interactivity, embodiment, and persistence. These features enable users to engage with a more immersive and interactive digital environment, providing a realistic and captivating experience. Hugues et al. (2011) argue that what the human augments is not reality but the perception of reality. They define reality as everything that exists and then claim that based on the reality definition, it is impossible to augment anything else to reality as it is already named everything, and based on the above description, they proposed augmented perception and artificial environment. Therefore, by clarifying the way of presence in this space, the concept of Avatar arises as a representation and alternative egos of users in this developed Virtual reality environment to do their daily jobs and activities (Kye et al., 2021).

LaValle defines virtual reality environment by using four key elements: targeted behavior, organism, artificial sensory stimulation, and awareness. In simpler terms, VR involves stimulating the senses to influence behavior in an organism, even if the organism is unaware of the influence (LaValle, 2023). VR stands out from other media due to its immersion and presence, which are not solely dependent on technology, according to Riva et al. (2003). Lombard and Ditton (1997) offer a unified definition of presence as the perception of an unmediated experience. Immersion pertains to how individuals perceive their interaction with VR and environment, rather than their physical surroundings, as described by Davis et al. (2009) and Morganson et al. (2015). Schuemie et al. (2001) identify key attributes of immersive media, including sensory and symbolic cues integration, interactivity, freedom to explore, believability, and plausibility. The main distinction between immersion and presence lies in their objective and subjective nature (Rauschnabel et al., 2022). Slater and Wilbur (1997) argue that immersion in virtual reality can be measured by five factors: inclusivity or physical reality exclusion, panoramic display to imply a broad view, sensory variety range and modalities accommodated, and vividness that includes resolution and fidelity. The output tries to make an immersive virtual reality or IVR.

Based on the consensus among scholars regarding the definition of the metaverse concept, its derivatives, and types, as well as the terms used in virtual environments such as immersion and presence, which are explained earlier, and technical requirements and capabilities, VR has been categorized into three main segments: non-immersive VR, semi-immersive VR, and fully immersive VR. The solutions developed for transferring virtual content to users are a balance of two components: the level of invasion and the level of immersion. Solutions that support more invasion, such as head-mounted displays (HMDs), provide a lower level of immersion for the user than the CAVE automatic virtual environment that follows the increased immersion (Carrozzino & Bergamasco, 2010). Most scholars are not aligned with Carrozzino and Bergamasco's notion, which, to some extent, the constant advancement and diversity of HMDs confirms this. Systems like CAVE can be classified as semi-immersive, and desktop-based VR is defined as non-immersive. To capture and represent objects that are far away and also large-scale, like walking through nature and outdoors, spherical panoramic 360-degree VR was developed to provide a more realistic essence but, in contrast, less in user interactions (Slater & Sanchez-Vives, 2016). It will delve into these requirements during the section that covers user and customer experience.

Samples of the Metaverse-Driven Tourism

Regardless of the categories announced in the above description, a quadruple category of metaverse can also be imagined from the perspective of the type of use. It is important to note that services on these platforms often overlap, making precise categorization challenging. Platforms can belong to multiple categories. Game-based metaverse offers interactive gameplay, quests, and game-like activities. Examples include Roblox, Decentral Games, Minecraft (with modding), Decentraland, Fortnite Creative, PokemonGo, Axie Infinity, Uplift World, Second Life, VRChat, The Sandbox, AltspaceVR, and Rec Room. Social metaverse emphasizes social interaction and virtual gatherings, allowing people to connect and communicate. Examples include VRChat, AltspaceVR, and Meta Quest Horizon. Blockchain-based metaverse utilizes blockchain technology to enable ownership and trade of digital assets like non-fungible tokens (NFTs), ensuring true ownership. Examples include Decentraland, The Sandbox, Somnium Space, CryptoVoxels, Axie Infinity, and The Uplift World. Enterprise Metaverse is tailored for professional activities, fostering collaboration, training, education, and work-related tasks within a digital space.

In the context of metaverse-driven tourism, examples of services and businesses that have embraced this technology include cyber tourism destinations, mentioning that most of these services are neither well-known nor user-friendly. TUI Group offers a comprehensive range of tourism services on a blockchain platform, as well as Budweiser and Stella Artois's Zed Run, Marriott Bonvoy, new Human Resource Management by STRIVR requested by MGM Resort Group, Metaverse Seoul, Infinity Moving Map (Southwest Airlines), Yucatan (Chichen Itza), Leven (a metaverse-driven virtual hotel), MoyaLand, Histovery, HistoPAD, ZEPETO World, and Hegra (an ancient city in KSA in the Metaverse).

Experience, User Experience, and Customer Experience

Experience plays a vital role in consumers' lives (Carù & Cova, 2003). Holbrook and Hirschman (1982) provided one of the insightful perspectives on the term experience in the early 1980s. They highlight its pleasurable and emotional attributes and emphasize consumption's sensory and imaginative elements. It is the world that Donald Norman describes in his book "Invisible Computer" too. A world in which digital technologies are constantly moving toward human life. New ideas as digital entities became an integral and fundamental element of human life to provide better services and performance (D. A. Norman, 1998). User-centric design as one of the main areas in ergonomics is a process in which product users' needs, desires, abilities, and limitations are focused on at all design stages, and product features are shaped accordingly (Sadeghi Naeini et al., 2022). Standards help to frame users' interactions. To shape merit experiences, In the late 20th century, ISO Standard 13407, provided four crucial steps for each user-centric project; identify the usage environment and understand the usage conditions, identify the user group and their needs, ideation, and evaluation based on user feedback (Maguire, 2001). Customer experience can be defined as the point of aggregation of all cognitive, emotional, behavioral, sensory, and social interactions across the entire customer journey with a company (Lemon & Verhoef, 2016). The scope and aim of the user-centric design approach is making experience, which prioritizes continuous refinement and improvement based on invaluable end-user feedback and specific requirements. Hassenzahl and Tractinsky (2006) also explained the user experience as a consequence of a user's internal state, such as expectations, needs, and motivations, the characteristics of the designed system, such as complexity, usability, and functionality, and the environment wherein the interaction takes place. Modern technologies aim to provide user-friendly experiences by emphasizing designed interactions in our daily lives. This will shift focus from mere functionality to practical engagement (Koleini Mamaghani & Morshedzadeh, 2011). User experience is defined simply as users' emotions, attitudes, thoughts, behaviors, and perceptions across the usage lifecycle (Hinderks et al., 2019). Becker and Jaakkola (2020) argued that customer experiences are spontaneous responses to stimuli rather than deliberate. Moreover, Bowman and colleagues have especially outlined three pivotal interaction domains in immersive virtual reality that demand research attention in design: navigation, manipulation, and system control (2001). It is important to understand that modifying or redesigning a system has far-reaching effects on its users. Therefore, it is necessary to consider and recognize the various sensory, cognitive, emotional, social, behavioral, spiritual, and moral experiences of users, as they all play an important role (Dahlgaard et al., 2008). It occurs before, during, and after users interact with services or products. Sharma et al. (2017) mentioned that augmented reality, virtual reality, and mixed reality are immersive technologies that allow interaction designers to create visually enhanced user experiences by merging the physical world with the virtual one. This opens up new opportunities for designers to facilitate the development of innovative and engaging user experiences. The other user experience method, "The Wheel," was proposed in four steps by Hartson and Pyla (2019) to overcome a critical issue due to the lack of a systematic approach to design. The issue affects the evaluation, usability enhancement, product efficiency, effectiveness, and, ultimately, user satisfaction. "The Wheel" encompasses comprehending requirements, designing solutions, prototyping, and user experience assessment. Popescu suggests that a mostly 2D interface can be more effective than an all-3D interface in some cases, such as when dealing with spreadsheets that require good readability. Therefore, it is more appropriate to use the term "the 3D-enhanced web" instead of "the 3D web" (Popescu, 2020). Afterward and recently, Nguyen et al. (2022) also proposed a framework and, in its complementary step, a methodology called DEMAND with a collaborative environment by collecting feedback from stakeholders and significant end-users to respond to user experience challenges in virtual reality and Industry 4.0 technology.

Besides the term user experience, affordance, which means to afford something, is the other crucial term. The idea of affordance was first proposed by cognitive psychologist James Gibson (2014) and focused on what the environment offers to living things, regardless of good or bad, what it provides or makes available. The term "functional possibilities" that Gibson proposed refers to the range of capabilities that the environment provides to the individuals involved. According to his research, affordance is influenced by environment and living factors, regardless of perception or attention. In the field of explaining user interactions and acquisition, Donald Norman proposed the use of affordance, which also has a notable role in human–computer interaction. Of course, it was slightly different from the original meaning presented by Gibson. Norman's emphasis was on perceived or real affordance. McGrenere and Ho (2000) also discuss the concept of affordance taxonomy and the functional hierarchy. Norman (2013) argues that affordances describe the relationship between a user and a product rather than a property of the product itself, as the existence of an affordance depends on the properties of both the product and the user. Hartson (2003) classified affordances into four types: cognitive, physical, sensory, and functional. Ono and Watanabe later presented a revised classification including cognitive, physical, and functional interactions (Morshedzadeh et al., 2016).

Seeger described an interactive system based on the perception and response process, an environment with components of the user's demographic and psychographic characteristics, as well as the product's properties, such as technical, economic, and human-related (Wartzack et al., 2019). The feedback between the user and the product is known as user–product interaction (Sadeghi Naeini, 2020; Wolf et al., 2019). The goal of interactive design is to minimize the user's challenges and simplify their experience while using the product, without reducing its practical and functional values. Numerous studies have consistently shown that technologies under the metaverse offer a notably more immersive experience than conventional counterparts, particularly in augmenting factors related to immersion, especially within the tourism domain. Immersive experiences entail multi-modal interaction, where a user is immersed in a digital space or digital elements integrate with the physical world, which can invoke a sense of presence. This fusion of immersion, immediacy, and presence is pivotal for a complete understanding of immersive media (Witmer & Singer, 1998). Perkis and colleagues (2020) define immersive media experience as a highly realistic simulation made by various senses and semiotic channels users encounter in a technology-driven setting to enable them to actively engage with the information and experiences offered by this simulated world. These technologies have been found to impact critical aspects such as heightened physical and cognitive engagement positively, amplified emotional involvement, increased enjoyment, and heightened sense of presence, as well as improvements in navigation and exploration (Hu et al., 2016; Patrick et al., 2000; Pollalis et al., 2017; Sousa Santos et al., 2009). The virtual tourists (as users) experience encompasses all the interactions, emotions, and memories that a tourist accumulates during their journey. It's not just about the physical places visited but also about the feelings, sensations, and knowledge gained during the trip. This includes the sights they see, the activities they engage in, the people they meet, the food they eat, and the overall atmosphere and ambiance of the destination. A positive experience often leads to higher satisfaction and a greater likelihood of return visits or positive word-of-mouth recommendations (Slater & Sanchez-Vives, 2016). The heart and soul of the tourism industry lies in the experiences it offers, and the tourism industry thrives on experiences (Kim & So, 2022). Tourist experiences can be divided into these segments: pre-trip expectations, physical or virtual environment, cultural and social interactions, service quality, and personal interpretation and perception.

With the advancement of network and mobile infrastructure and individuals staying home instead of venturing out to explore destinations, digital storytelling and intricacies of experience design have gained prominence because, typically, the excitement of certain tangible and real-world experiences can never be fully replicated by a virtual one. Undoubtedly, the impact of the COVID-19 pandemic cannot be understated, as it heightened and solidified this requirement, hastening the sense of telepresence adoption.

Tourists' Decision-Making

Understanding the users' journey from choosing a metaverse-driven tourism platform or VR content to becoming their VR experiences is essential as same as the travelers' and tourists' decision-making processes in choosing the destination in the real world. The use of fully immersive virtual reality in tourism marketing is a popular topic of discussion. Of course, traditional travel guides are still preferred (Tussyadiah et al., 2018). It can facilitate the decision-making process toward the destination by stimulating users' and potential tourists' interest, curiosity, and willingness (Beck & Egger, 2018).

Virtual reality holds the potential to enhance the inspiration and information stages of the customer journey by offering tourists a chance to get a preview of the tourism experience and interact with reliable and comprehensive information (Tussyadiah et al., 2018). The VR/AR effect is not limited to tourists' on-site experience and is detectable and traceable in the post-travel phase (Marasco et al., 2018). Such immersive virtual environments have the potential to aid in users' and tourists' decision-making by enhancing sensory experiences (Raptis et al., 2018; Stevens & Kincaid, 2015).

Sustainable Development Goals

Rising population and imbalanced consumption strain societies, causing environmental and social tensions. The 2030 Sustainable Development Goals (SDGs) built on the earlier Millennium Development Goals and integrated the International Labor Organization's 1999 Decent Work Agenda, emphasizing job promotion, labor rights, social protection, and gender equality (Pedersen, 2018; Winchenbach et al., 2019). Goal 8 of the SDGs centers on "Workplace Dignity" (Hodson, 2001). Interpretations of sustainability and related terms vary, especially in different ideological contexts. However, with all the drawbacks, achieving a consensus on goals minimally may be preferable to adopting potentially conflicting strategies.

The tourism industry, while facing a decline post-pandemic, remains a major global employer. It creates one in 11 new jobs worldwide, employing 289 million people and contributing 6.1% to the global gross domestic product (GDP) (WTTC, 2022). Tourism jobs cover a wide range, from online platforms to physical locations. While they offer benefits, there are concerns for job seekers in this sector. Of course, The UN's International Year of Sustainable Tourism for Development (IY2017) has rehighlighted its role in sustainable progress.

Dignity violations at work stem from power imbalances, encompassing financial, educational, and cultural disparities. This disproportionately affects lower-tier employees in precarious situations, making dignified employment harder to attain (Jacobson, 2009). This aligns with concerns highlighted by the ILO and SDGs. Companies face challenges due to unstable policies and unexpected events like the pandemic, while tourism jobs are often temporary, jeopardizing job security. Environmental activists also raise valid concerns about the negative impacts of tourism. Implementing high-level political agreements like the SDGs into the market is difficult, especially for short-term-focused businesses (Pedersen, 2018). This underscores the importance of corporate social responsibility (CSR) in guiding initiatives and practices. Developing a metaverse tourism platform demands substantial technical resources and support for the R&D team, either internally or via open innovation partnerships. Concerns about resource allocation might impede progress. Hristov and Chirico (2019) suggest that adopting a CSR approach can boost reputation and goodwill. Companies can showcase their dedication to climate change by investing in metaverse-driven tourism, which includes efforts to combat air travel pollution and preserve historical sites. This approach supports smooth transitions with technical changes, reducing risks. Companies have aligned performance metrics with societal benefits, driven by SDGs, especially Goal 8, which focuses on decent work (Williams et al., 2019). However, current tourism policies may not fully align with this trend and may not meet public demand. Blockchain technology, known for decentralization, democratization, and enhanced security, can benefit metaverse-driven tourism platforms.

Quality of Experience

Consider a digital traveler in the metaverse during the trip phase who purchases a handicraft work or keepsake as a digital asset from a platform as a scenario. How would one perceive, comprehend, and assess the experience of this transaction? The experience made out of the process of exchanging digital assets as an asymmetric bond differs significantly from the user experience associated with physical products, which is a unique connection between two entities built on trust, and prior interaction accurately indicates the trust the trustor has in the trustee and will be evaluated by their feedback after each exchange (Tagliaferri & Aldini, 2018; Truong et al., 2018). The preceding section addressed with the notions of experience and the experiences of users and customers based on the research conducted by scholars. Given the interdisciplinary nature of this subject, exploring the concept of quality of experience (QoE) would be beneficial. To grasp the QoE term, it could be helpful to begin by dissecting and elucidating the terminology within its components. Quality encompasses the extent to how well a product, service, or process aligns with specified requirements and fulfills stated or implied needs. It is a subjective concept, subject to individual or sector-specific assessment and interpretation. In technical jargon, quality encompasses the inherent attributes that enable a product or service to effectively serve its intended function. Organizations such as ISO, ASQ, and EFQ offer defined frameworks for quality management, underscoring the significance of fitness for use, satisfying customer expectations, optimizing performance, and attaining enduring success through a balanced approach (Akhtar & Falk, 2017; Martens & Martens, 2001; Möller & Raake, 2014). As Jekosch (2005) mentioned, quality forms in two paths: the reference path, which considers past experiences and the temporal nature of quality, and the perception path, which starts with a physical event processed through sensory and cognitive processes. These paths converge in the comparison and judgment process, leading to experienced quality, which is gathered descriptively and time-bound for each user. QoE has risen in importance since the last years of the 1990s. In communication and telecommunication, quality of service (QoS) is the focus, but QoE emerged, recognizing that QoS alone could not capture the complexities of modern services. While some studies may equate performance and QoE, they are distinct concepts. Performance focuses on delivering the promised function to the user, while QoE assesses the user's perception of both stated and unstated functions of a product or service.

QoE phenomenon refers to the user's level of satisfaction or frustration with an immersive media experience in an application or service. The QoE unquestionably overlaps with user experience and is impacted by human, system, and contextual factors, which are named influence factors. Experienced quality results from a cognitive process of comparing and judging perceived quality attributes against desired features based on user expectations, all prompted by a physical signal (Brunnström et al., 2013). QoE refers to the overall measure of user satisfaction with the performance of a service, application, or product. It can be considered a comprehensive key performance indicator that encompasses various factors such as responsiveness, reliability, usability, and efficiency. QoE features can be identified perceptually and interpretively, enabling multidimensional analysis and qualitative assessments aided by regression techniques in creating QoE feature sets to establish their importance in quality preferences (S. Egger et al., 2012). Categorizing QoE features is crucial for evaluation. They can be classified into four levels: direct perception, which is considered a temporal dimension too, includes sensory factors; level of interaction, which involves responsiveness and communication efficiency; usage situation, which covers aspects such as accessibility and stability; and service level, encompassing aesthetics, usability, and long-term satisfaction (Möller & Raake, 2014). The Mean Opinion Score rating test, recommended by the International Telecommunication Union, is a widely acknowledged framework for assessing QoE; additionally, other valuable assessment frameworks include Absolute Category Rating, Double Stimulus Continuous Quality Evaluation, and Video Quality Metric (K.-T. Chen et al., 2009; Recommendation, 2008).

Conclusion, Limitations, and Future Research

QoEs can be described as the relationship among the metaverse service provider, the network infrastructure provider, and the users as vertices of the Human, System, and Context triangle. Perceptive and interpretive identification of QoE features allows for multidimensional analysis and qualitative assessments in crafting feature sets to determine their significance in quality preferences. Metaverse-driven tourism impacts various aspects. As humans are one of the impact factors affected by metaverse-driven tourism, users can be classified into significant classification of suppliers and tourists. Suppliers gain from the metaverse, although they face complexity. Still, through transformative operations, they benefit from a more creative and productive workforce (without being hindered by factors like disability, age, or gender). Of course, the idea that ICT possesses transformative power is not a recent concept, as noted by Zuboff (1988). As a segment of suppliers, tourism managers should leverage this concept and the wealth of data for tasks like creating and designing new online products, promotions, and destination management organization materials, as well as strategic decision-making and customized and updated CRM, to follow up on sustainable goals. This transforms trip planning for tourists, granting instant access to their desired destinations and information through 3D immersive virtual experiences, while social media is integrated into the proposed service as default to share real-time travel experiences with their other friends in avatar format. Enhancement of the coordination among two other influential impact factors and seamless collaboration to make an optimal user experience is crucial. Multi-sensory incorporation of olfactory and haptic feedback aims to improve the presence and interaction and reduce ambiguity in the duality of virtuality reality, along with visional and auditory modalities, which had been included in VR experiences. As Vlachos (2013) categorized organizations involved in Metaverse-driven tourism businesses into five adopter categories, leaders and tech experts have to accelerate metaverse-driven tourism development to encourage the rest of the activists, such as small and medium enterprises and local businesses that adopt late. For instance, educational centers, alongside other businesses and service providers, should try to train this ecosystem's capable and skilled forces and increase users' public technology literacy for broader acceptance. In the near future, healthier lifestyles will lead to extended life expectancy for the elderly (Siegelman, 2014). As technology advances, it presents unique needs and hurdles for older users. This creates an opening for tailored learning solutions within the metaverse, aiming to enhance creativity and independence for this demographic. Factors such as emotional well-being, tech comfort, training effectiveness, self-confidence, adherence, motivation, and cognitive skills should be focal points in this design effort. Paying attention to the demographics of Generation Z and Alpha is also necessary due to their needs and requirements.

Attaining and realizing a completely immersive user experience aligned with the emerging and burgeoning notion of metaverse-driven tourism, characterized by its multi-dimensional nature and diverse technical prerequisites, cannot only be achieved through the careful implementation of Web 2.0 realm standards and criteria essential for delivering services. The lack of enough access to XR accessories for users to have a palpable presence and utilization of the offered services and also the limited number of developed instances. Additionally, the constructed metaverses remained disconnected from one another, resulting in these nascent realms being solitary and insulated despite their considerable potential. Furthermore, due to the interdisciplinary nature of research and the multiplicity and diversity of resources in various fields, the other crucial constraint of this research lies in the lack of enough empirical substantiation. After that, the theoretical foundation of its outcome has not been translated into policy and into practice yet. In other words, the conceptualization of providing tourism services within a metaverse framework was still in its early stages. Naturally, given the extensive array of sources, gathering every opinion proved unfeasible. Moreover, engaging in discussions with the expert group proved highly beneficial in gaining a clearer path forward.

Funding: This research received no external funding.
Conflicts of Interest: The authors declare no conflict of interest.

References

Akhtar, Z., & Falk, T. H. (2017). Audio-visual multimedia quality assessment: A comprehensive survey. *IEEE Access, 5*, 21090–21117.

Ameen, N., Cheah, J.-H., & Kumar, S. (2022). It's all part of the customer journey: The impact of augmented reality, chatbots, and social media on the body image and self-esteem of Generation Z female consumers. *Psychology & Marketing, 39*(11), 2110–2129.

Beck, J., & Egger, R. (2018). Emotionalise me: Self-reporting and arousal measurements in virtual tourism environments. *Information and Communication Technologies in Tourism 2018: Proceedings of the International Conference in Jönköping, Sweden, January 24–26, 2018*, 3–15.

Becker, L., & Jaakkola, E. (2020). Customer experience: Fundamental premises and implications for research. *Journal of the Academy of Marketing Science, 48*, 630–648.

Bobrowsky, M., & Needleman, S. (2022). *What is the metaverse? The future vision for the internet*. Retrieved August 25, 2022, from https://www.wsj.com/story/what-is-the-metaverse-the-future-vision-for-the-internet-ca97bd98.

Bowman, D. A., Kruijff, E., LaViola, J. J., & Poupyrev, I. (2001). An introduction to 3-D user interface design. *Presence, 10*(1), 96–108.

Brunnström, K., Beker, S. A., De Moor, K., Dooms, A., Egger, S., Garcia, M.-N., Hossfeld, T., Jumisko-Pyykkö, S., Keimel, C., & Larabi, M.-C. (2013). *Qualinet white paper on definitions of quality of experience*. https://hal.science/hal-00977812.

Budeanu, A., Miller, G., Moscardo, G., & Ooi, C.-S. (2016). Sustainable tourism, progress, challenges and opportunities: An introduction. In *Journal of cleaner production* (Vol. 111, pp. 285–294). Elsevier.

Carrozzino, M., & Bergamasco, M. (2010). Beyond virtual museums: Experiencing immersive virtual reality in real museums. *Journal of Cultural Heritage, 11*(4), 452–458.

Carù, A., & Cova, B. (2003). Revisiting consumption experience: A more humble but complete view of the concept. *Marketing Theory, 3*(2), 267–286.

Chen, K.-T., Wu, C.-C., Chang, Y.-C., & Lei, C.-L. (2009). A crowdsourceable QoE evaluation framework for multimedia content. *Proceedings of the 17th ACM International Conference on Multimedia*, 491–500.

Chen, S., Tian, D., Law, R., & Zhang, M. (2022). Bibliometric and visualized review of smart tourism research. *International Journal of Tourism Research, 24*(2), 298–307.

Cheong, R. (1995). The virtual threat to travel and tourism. *Tourism Management, 16*(6), 417–422.

Choi, H., & Kim, S. (2017). A content service deployment plan for metaverse museum exhibitions – Centering on the combination of beacons and HMDs. *International Journal of Information Management, 37*(1), 1519–1527.

Culot, G., Nassimbeni, G., Orzes, G., & Sartor, M. (2020). Behind the definition of Industry 4.0: Analysis and open questions. *International Journal of Production Economics, 226*, 107617.

Dahlgaard, J. J., Khanji, G. K., & Kristensen, K. (2008). *Fundamentals of total quality management*. Routledge. https://books.google.com/books?hl=en&lr=&id=kuSQAgAAQBAJ&oi=fnd&pg=PP1&dq=Dahlgaard+et+al.,+2008&ots=dLWcr6L7N8&sig=AVUl8pmCdfU-6y49PFM-zHYKVP0

Davis, A., Murphy, J., Owens, D., Khazanchi, D., & Zigurs, I. (2009). Avatars, people, and virtual worlds: Foundations for research in metaverses. *Journal of the Association for Information Systems*, *10*(2), 1.

Desai, P. R., Desai, P. N., Ajmera, K. D., & Mehta, K. (2014). A review paper on oculus rift-a virtual reality headset. *arXiv Preprint*, arXiv:1408.1173.

Dolnicar, S., Xiang, Z., Fuchs, M., Gretzel, U., & Höpken, W. (2020). *Handbook of e-Tourism*. Springer.

Duan, H., Li, J., Fan, S., Lin, Z., Wu, X., & Cai, W. (2021). Metaverse for social good: A university campus prototype. *Proceedings of the 29th ACM International Conference on Multimedia*, 153–161.

Egger, I., Lei, S. I., & Wassler, P. (2020). Digital free tourism – An exploratory study of tourist motivations. *Tourism Management*, *79*, 104098.

Egger, S., Reichl, P., Hoßfeld, T., & Schatz, R. (2012). "Time is bandwidth"? Narrowing the gap between subjective time perception and quality of experience. *2012 IEEE International Conference on Communications (ICC)*, 1325–1330.

Farshid, M., Paschen, J., Eriksson, T., & Kietzmann, J. (2018). Go boldly! Explore augmented reality (AR), virtual reality (VR), and mixed reality (MR) for business. *Business Horizons*, *61*(5), 657–663.

Filieri, R., D'Amico, E., Destefanis, A., Paolucci, E., & Raguseo, E. (2021). Artificial intelligence (AI) for tourism: An European-based study on successful AI tourism start-ups. *International Journal of Contemporary Hospitality Management*, *33*(11), 4099–4125.

Gibson, J. J. (2014). *The ecological approach to visual perception: Classic edition*. Psychology Press.

Go, H., & Gretzel, U. (2009). Web 3.0: Tourism in virtual worlds. *Proceedings of the Hospitality and Information Technology Association (HITA). 2009 Conference, Anaheim, CA, June 21–22*.

Gong, C., & Ribiere, V. (2021). Developing a unified definition of digital transformation. *Technovation*, *102*, 102217.

Gursoy, D., Malodia, S., & Dhir, A. (2022). The metaverse in the hospitality and tourism industry: An overview of current trends and future research directions. *Journal of Hospitality Marketing & Management*, *31*(5), 527–534.

Guttentag, D. A. (2010). Virtual reality: Applications and implications for tourism. *Tourism Management*, *31*(5), 637–651.

Hartl, E., & Hess, T. (2019). IT projects in digital transformation: A socio-technical journey towards technochange. In *Proceedings of the 27th European Conference on Information Systems (ECIS), Stockholm, Uppsala, Sweden, June 8–14, 2019*. ISBN 978-1-7336325-0-8.

Hartson, R. (2003). Cognitive, physical, sensory, and functional affordances in interaction design. *Behaviour & Information Technology*, *22*(5), 315–338. https://doi.org/10.1080/01449290310001592587

Hartson, R., & Pyla, P. (2019). Agile lifecycle processes and the funnel model of agile UX. *The UX Book*, 63–80.

Hassenzahl, M., & Tractinsky, N. (2006). User experience-a research agenda. *Behaviour & Information Technology*, *25*(2), 91–97.

Hinderks, A., Schrepp, M., Mayo, F. J. D., Escalona, M. J., & Thomaschewski, J. (2019). Developing a UX KPI based on the user experience questionnaire. *Computer Standards & Interfaces*, *65*, 38–44.

Hodson, R. (2001). *Dignity at work*. Cambridge University Press. https://books.google.com/books?hl=en&lr=&id=HwVKKFDfIeMC&oi=fnd&pg=PR10&dq=Hodson,+R.+

(2001).+Dignity+at+work.+Cambridge,+UK:+Cambridge+University+Press.&ots=Q9g yokZmhY&sig=z5PiWfT_ArQ5hnljESDMaVZ5Qys

Holbrook, M. B., & Hirschman, E. C. (1982). The experiential aspects of consumption: Consumer fantasies, feelings, and fun. *Journal of Consumer Research, 9*(2), 132–140.

Hristov, I., & Chirico, A. (2019). The role of sustainability key performance indicators (KPIs) in implementing sustainable strategies. *Sustainability, 11*(20), 5742.

Hu, G., Bin Hannan, N., Tearo, K., Bastos, A., & Reilly, D. (2016). Doing while thinking: Physical and cognitive engagement and immersion in mixed reality games. *Proceedings of the 2016 ACM Conference on Designing Interactive Systems*, 947–958.

Hugues, O., Fuchs, P., & Nannipieri, O. (2011). New augmented reality taxonomy: Technologies and features of augmented environment. *Handbook of Augmented Reality*, 47–63.

Jacobson, N. (2009). A taxonomy of dignity: A grounded theory study. *BMC International Health and Human Rights, 9*(1), 3. https://doi.org/10.1186/1472-698X-9-3

Jekosch, U. (2005). *Voice and speech quality perception: Assessment and evaluation.* Springer Science & Business Media.

Kim, H., & So, K. K. F. (2022). Two decades of customer experience research in hospitality and tourism: A bibliometric analysis and thematic content analysis. *International Journal of Hospitality Management, 100*, 103082.

Koleini Mamaghani, N., & Morshedzadeh, E. (2011). Evaluation of user and product's function using interaction design method. *Honar-Ha-Ye-Ziba: Honar-Ha-Ye-Tajassomi, 2*(41), 95–104.

Kunkel, N., & Soechtig, S. (2017). Mixed reality: Experiences get more intuitive, immersive and empowering. In *Tech trends 2017. The kinetic enterprise.* Deloitte.

Kwon, J., & Koo, C. (2022). TechTalk with the editors of journal of travel research: The perspectives of technology by Dr. Nancy G. McGehee and Dr. James Petrick. *Journal of Smart Tourism, 2*(1), 1–3.

Kye, B., Han, N., Kim, E., Park, Y., & Jo, S. (2021). Educational applications of metaverse: Possibilities and limitations. *Journal of Educational Evaluation for Health Professions, 18*, 32. https://doi.org/10.3352/jeehp.2021.18.32

LaValle, S. M. (2023). *Virtual reality.* Cambridge University Press.

Lee, L.-H., Braud, T., Zhou, P., Wang, L., Xu, D., Lin, Z., Kumar, A., Bermejo, C., & Hui, P. (2021). All one needs to know about metaverse: A complete survey on technological singularity, virtual ecosystem, and research agenda. *arXiv Preprint*, arXiv:2110.05352.

Leiper, N. (1979). The framework of tourism: Towards a definition of tourism, tourist, and the tourist industry. *Annals of Tourism Research, 6*(4), 390–407.

Lemon, K. N., & Verhoef, P. C. (2016). Understanding customer experience throughout the customer journey. *Journal of Marketing, 80*(6), 69–96.

Li, J., Xu, L., Tang, L., Wang, S., & Li, L. (2018). Big data in tourism research: A literature review. *Tourism Management, 68*, 301–323.

Li, Y., Hu, C., Huang, C., & Duan, L. (2017). The concept of smart tourism in the context of tourism information services. *Tourism Management, 58*, 293–300.

Liu, J., Wang, C., Zhang, T., & Qiao, H. (2023). Delineating the effects of social media marketing activities on Generation Z travel behaviors. *Journal of Travel Research, 62*(5), 1140–1158.

Lombard, M., & Ditton, T. (1997). At the heart of it all: The concept of presence. *Journal of Computer-Mediated Communication, 3*(2), JCMC321.

Maguire, M. (2001). Methods to support human-centred design. *International Journal of Human-Computer Studies, 55*(4), 587–634.

Maha, A., Donici, A. N., & Postolachi, A. T. (2012). Electronic tourism (E-tourism)-a theoretical approach. *MPRA Paper 41745*, University Library of Munich, Germany.

Makransky, G., & Mayer, R. E. (2022). Benefits of taking a virtual field trip in immersive virtual reality: Evidence for the immersion principle in multimedia learning. *Educational Psychology Review, 34*(3), 1771–1798. https://doi.org/10.1007/s10648-022-09675-4

Marasco, A., Buonincontri, P., Van Niekerk, M., Orlowski, M., & Okumus, F. (2018). Exploring the role of next-generation virtual technologies in destination marketing. *Journal of Destination Marketing & Management, 9*, 138–148.

Martens, H., & Martens, M. (2001). *Multivariate analysis of quality: An introduction.* John Wiley & Sons.

McGrenere, J., & Ho, W. (2000). Affordances: Clarifying and evolving a concept. *Graphics Interface, 2000*, 179–186. http://graphicsinterface.org/wp-content/uploads/gi2000-24.pdf

McIntosh, R. W., Goeldner, C. R., & Ritchie, J. B. (1995). *Tourism: Principles, practices, philosophies* (Issue 7th ed.). John Wiley and Sons.

Milgram, P., & Kishino, F. (1994). A taxonomy of mixed reality visual displays. *IEICE Transactions on Information and Systems, 77*(12), 1321–1329.

Milgram, P., Takemura, H., Utsumi, A., & Kishino, F. (1995). Augmented reality: A class of displays on the reality-virtuality continuum. *Telemanipulator and Telepresence Technologies, 2351*, 282–292.

Möller, S., & Raake, A. (2014). *Quality of experience: Advanced concepts, applications and methods.* Springer.

Morganson, V. J., Major, D. A., Streets, V. N., Litano, M. L., & Myers, D. P. (2015). Using embeddedness theory to understand and promote persistence in stem majors. *The Career Development Quarterly, 63*(4), 348–362. https://doi.org/10.1002/cdq.12033

Morshedzadeh, E., Ono, K., & Watanabe, M. (2016). A new model for improving user-product interaction evaluation, based on affordance and factor analysis. *Bulletin of Japanese Society for the Science of Design, 62*(5), 5_41–5_48.

Mura, P., Tavakoli, R., & Pahlevan Sharif, S. (2017). "Authentic but not too much": Exploring perceptions of authenticity of virtual tourism. *Information Technology & Tourism, 17*(2), 145–159.

Nguyen, H. N., Lasa, G., Iriarte, I., Atxa, A., Unamuno, G., & Galfarsoro, G. (2022). Human-centered design for advanced services: A multidimensional design methodology. *Advanced Engineering Informatics, 53*, 101720. https://doi.org/10.1016/j.aei.2022.101720

Norman, D. (2013). *The design of everyday things: Revised and expanded edition.* Basic Books.

Norman, D. A. (1998). *The invisible computer: Why good products can fail, the personal computer is so complex, and information appliances are the solution.* MIT Press.

O'Sullivan, M. J., & Grigoras, D. (2015). Integrating mobile and cloud resources management using the cloud personal assistant. *Simulation Modelling Practice and Theory, 50*, 20–41.

Patrick, E., Cosgrove, D., Slavkovic, A., Rode, J. A., Verratti, T., & Chiselko, G. (2000). Using a large projection screen as an alternative to head-mounted displays for virtual environments. *Proceedings of the SIGCHI Conference on Human Factors in Computing Systems*, 478–485.

Pedersen, C. S. (2018). The UN sustainable development goals (SDGs) are a great gift to business! *Procedia CIRP, 69*, 21–24.

Perkis, A., Timmerer, C., Baraković, S., Husić, J. B., Bech, S., Bosse, S., Botev, J., Brunnström, K., Cruz, L., & De Moor, K. (2020). QUALINET white paper on definitions of immersive media experience (IMEx). *arXiv Preprint*, arXiv:2007.07032.

Perry Hobson, J. S., & Williams, A. P. (1995). Virtual reality: A new horizon for the tourism industry. *Journal of Vacation Marketing, 1*(2), 124–135.

Pollalis, C., Fahnbulleh, W., Tynes, J., & Shaer, O. (2017). HoloMuse: Enhancing engagement with archaeological artifacts through gesture-based interaction with holograms. *Proceedings of the Eleventh International Conference on Tangible, Embedded, and Embodied Interaction*, 565–570.

Popescu, A.-D. (2020). Decentralized finance (defi)–the Lego of finance. *Social Sciences and Education Research Review, 7*(1), 321–349.

Rajaee, F. (2000). *Globalization on trial: The human condition and the information civilization.* IDRC.

Raptis, G. E., Fidas, C., & Avouris, N. (2018). Effects of mixed-reality on players' behaviour and immersion in a cultural tourism game: A cognitive processing perspective. *International Journal of Human-Computer Studies, 114*, 69–79.

Rauschnabel, P. A., Felix, R., Hinsch, C., Shahab, H., & Alt, F. (2022). What is XR? Towards a framework for augmented and virtual reality. *Computers in Human Behavior, 133*, 107289.

Recommendation, I. (2008, August). *247 Objective perceptual multimedia video quality measurement in the presence of a full reference.* Retrieved August 2008, from https://www.itu.int/rec/T-REC-J.247-200808-I/en.

Riva, G., Davide, F., & IJsselsteijn, W. A. (2003). Being there: The experience of presence in mediated environments. *Being There: Concepts, Effects and Measurement of User Presence in Synthetic Environments, 5.*

Rocha, Á., Adeli, H., Reis, L. P., & Costanzo, S. (2018). *Trends and advances in information systems and technologies* (Vol. 1). Springer. https://doi.org/10.1007/978-3-319-77703-0.

Sadeghi Naeini, H. (2020). Ergonomics on the context of sustainability: A new approach on quality of life. *Iran University of Science & Technology, 30*(2), 260–271.

Sadeghi Naeini, H., Maria Conti, G., & Mosaddad, S. H. (2022). Industrial design evolution in the context of ergonomics and Industry 5.0. *Journal of Design Thinking, 3*(2). https://journals.ut.ac.ir/article_91021.html

Scholz, T. M., & Vyugina, D. (2019). Looking into the future: What we are expecting from generation Z. In *Generations Z in Europe* (pp. 277–284). Emerald Publishing Limited.

Schuemie, M. J., Van Der Straaten, P., Krijn, M., & Van Der Mast, C. A. (2001). Research on presence in virtual reality: A survey. *Cyberpsychology & Behavior, 4*(2), 183–201.

Shafiee, S., Ghatari, A. R., Hasanzadeh, A., & Jahanyan, S. (2019). Developing a model for sustainable smart tourism destinations: A systematic review. *Tourism Management Perspectives, 31*, 287–300.

Sharma, H. N., Alharthi, S. A., Dolgov, I., & Toups, Z. O. (2017). A framework supporting selecting space to make place in spatial mixed reality play. *Proceedings of the Annual Symposium on Computer-Human Interaction in Play*, 83–100.

Siegelman, P. (2014). *The compromised worker and the limits of employment discrimination law.* Available at SSRN 2485927. https://papers.ssrn.com/sol3/papers.cfm?abstract_id=2485927

Slater, M., & Sanchez-Vives, M. V. (2016). Enhancing our lives with immersive virtual reality. *Frontiers in Robotics and AI, 3*, 74.

Slater, M., & Wilbur, S. (1997). A framework for immersive virtual environments (FIVE): Speculations on the role of presence in virtual environments. *Presence: Teleoperators & Virtual Environments, 6*(6), 603–616.

Sousa Santos, B., Dias, P., Pimentel, A., Baggerman, J.-W., Ferreira, C., Silva, S., & Madeira, J. (2009). Head-mounted display versus desktop for 3D navigation in virtual reality: A user study. *Multimedia Tools and Applications, 41*, 161–181.

Stevens, J., & Kincaid, J. P. (2015). The effect of visual display on performance in mixed reality simulation. *International Journal of Modelling and Simulation, 35*(1), 20–26.

Tagliaferri, M., & Aldini, A. (2018). A taxonomy of computational models for trust computing in decision-making procedures. *European Conference on Cyber Warfare and Security*, 571–XVI. https://search.proquest.com/openview/cfdf62862eb5ba2f7d2eec2fff54054b/1?pq-origsite=gscholar&cbl=396497

Truong, N. B., Um, T.-W., Zhou, B., & Lee, G. M. (2018). Strengthening the blockchain-based internet of value with trust. *2018 IEEE International Conference on Communications (ICC)*, 1–7.

Tussyadiah, I. P., Wang, D., Jung, T. H., & Tom Dieck, M. C. (2018). Virtual reality, presence, and attitude change: Empirical evidence from tourism. *Tourism Management, 66,* 140–154.

Verhoef, P. C., Broekhuizen, T., Bart, Y., Bhattacharya, A., Dong, J. Q., Fabian, N., & Haenlein, M. (2021). Digital transformation: A multidisciplinary reflection and research agenda. *Journal of Business Research, 122,* 889–901.

Vlachos, I. (2013). Investigating e-business practices in tourism: A comparative analysis of three countries. *TOURISMOS: An International Multidisciplinary Refereed Journal of Tourism, 8*(1), 179–198.

Wartzack, S., Schröppel, T., Wolf, A., & Miehling, J. (2019). Roadmap to consider physiological and psychological aspects of user-product interactions in virtual product engineering. *Proceedings of the Design Society: International Conference on Engineering Design, 1*(1), 3989–3998.

Williams, A., Whiteman, G., & Parker, J. N. (2019). Backstage interorganizational collaboration: Corporate endorsement of the sustainable development goals. *Academy of Management Discoveries, 5*(4), 367–395.

Winchenbach, A., Hanna, P., & Miller, G. (2019). Rethinking decent work: The value of dignity in tourism employment. *Journal of Sustainable Tourism, 27*(7), 1026–1043. https://doi.org/10.1080/09669582.2019.1566346

Witmer, B. G., & Singer, M. J. (1998). Measuring presence in virtual environments: A presence questionnaire. *Presence, 7*(3), 225–240.

Wolf, A., Binder, N., Miehling, J., & Wartzack, S. (2019). Towards virtual assessment of human factors: A concept for data driven prediction and analysis of physical user-product interactions. *Proceedings of the Design Society: International Conference on Engineering Design, 1*(1), 4029–4038.

WTTC. (2022). *Travel & tourism could grow to $8.6 trillion in 2022, says WTTC.* Retrieved from https://wttc.org/News-Article/Travel-and-Tourism-could-grow-to-8-point-6-trillion-USD-in-2022-say-WTTC.

Zuboff, S. (1988). Dilemmas of transformation in the age of the smart machine. *Pub Type, 81.*

7 What Lies in the Current Scholarly Art of Work on Metaverse and Tourism? Exploring Future Research Directions

Manpreet Arora and Marco Valeri

Introduction

A collective virtual shared place that is generated when numerous virtual worlds or augmented reality environments come together is referred to as the "metaverse." A computer-generated environment and other users can be interacted with in this persistent, immersive, and vast digital reality. With the development of AR and VR technology, the concept of the metaverse gained popularity. It is frequently portrayed as an immersive, three-dimensional area where people may do a range of tasks, such as working, socializing, gaming, shopping, learning, and more. The metaverse has enormous potential. It has the potential to transform a number of fields and facets of our lives, such as business, education, communication, entertainment (Park & Kim, 2022; Valeri, 2023, 2024), and even remote work. By enabling global connections in a common virtual environment, it has the potential to completely transform social interactions by removing geographical boundaries. Many believe that the metaverse will be the next big thing in gaming, offering experiences that are more engaging and immersive than anything that is now feasible (Kye et al., 2021).

The potential for virtual real estate, digital assets, and a whole economy operating inside the metaverse could open up new business prospects. The metaverse may present novel opportunities for learning through simulations and immersive encounters. The metaverse does, however, also bring up several crucial issues, such as privacy, security (Wang et al., 2022), digital ownership, governance, and the possible effects on interpersonal relationships and mental health. As the metaverse develops further, norms, laws, and ethical issues will need to be addressed. These are continuous discussions. Large digital corporations are making investments in technology related to the metaverse (Kim, 2021); VR platforms, online games, and social media are examples of how this idea is being implemented in its early phases. It may take years for the metaverse to fully materialize, and it will probably change gradually as a result of user adoption and technological improvements. The metaverse offers fresh perspectives on and avenues for interaction inside virtual worlds, suggesting a possible progression in the way we engage with digital places. Due to its unique nature and potential benefits as well as challenges, it is attracting a lot of interest and investment from a variety of industries (Lee et al.,2021).

DOI: 10.4324/9781003497004-8

The metaverse has the power to profoundly affect a number of industries, opening up new avenues for customer interaction, business model development, and innovation. Businesses in a variety of sectors will need to modify and advance their plans in order to take advantage of the potential this new digital arena offers.

The Metaverse's Revolutionary Impact on Business

The idea of the metaverse has caused a paradigm shift in business, changing how many businesses function and engage with customers. This vast digital domain, this communal virtual arena, is changing how companies interact with their customers and add value (Arora & Sharma, 2023). Its impact extends to various industries, resulting in a surge of creativity, fresh prospects, and innovative difficulties. Modern business is about to undergo a multidimensional transformation due to the rise of the metaverse, a collective virtual shared place generated by the confluence of physical and virtual reality. After and during COVID-19, the industry, entrepreneurship, and every sector of the economy faced a huge shift and virtual world, entrepreneurial opportunities and digitalization assumed great importance in every respect (Arora & Sharma, 2021; Arora & Sharma, 2022a, 2022b). However, on the same hand, digitalization and excessive use of social media platforms posed challenges such as fake news, propaganda, and the spread of misinformation which has to be taken care of (Arora, 2020).

To begin with, the metaverse holds the potential to revolutionize conventional approaches to business and customer interaction. Companies are already experimenting with this immersive setting to build virtual experiences and stores. Businesses can engage with customers in new ways in the metaverse by providing immersive shopping experiences, virtual product trials, and demonstrations. This change broadens the market and provides a new level of individualized customer service, allowing businesses to collect priceless information on the tastes and actions of their customers. Furthermore, the metaverse overcomes the limitations of physical presence, allowing for hitherto unheard-of levels of remote work and cooperation. Companies can use this technology to create remote work environments where workers from all over the world can work together in a seamless manner, promoting efficiency and inclusivity. Reductions in the running expenses related to the upkeep of physical office buildings could also result from this change in the workplace. The metaverse also offers fresh possibilities for advertising and marketing tactics (Rather, 2023). Companies can design interactive, immersive marketing campaigns that incorporate their goods and services into this virtual environment. A significant change that has the potential to completely change the marketing landscape is the capacity to interact with customers more deeply and provide experiences rather than just commercials.

Moreover, the metaverse exhibits considerable potential for sectors apart from marketing and retail (Özdemir Uçgun & Şahin, 2023). With new learning opportunities, remote medical treatment, and immersive entertainment experiences, it has the potential to completely transform education, healthcare, and entertainment. The metaverse's interactive and collaborative features have the potential to revolution-ize the ways in which people study, get medical care, and pass the time.

However, there are also issues with the emergence of the metaverse. It is neces-sary to address privacy, security, and ethical issues related to data usage and virtual interactions. Ensuring user data privacy and creating a safe and secure workplace will be crucial as organizations negotiate this new terrain. The metaverse has the potential to completely rewrite the rules of contemporary business by revolution-izing marketing tactics (Buhalis et al., 2023), changing the way people interact with brands, and opening up new avenues for creativity in a variety of fields. It has a significant potential to innovate and disrupt the way business is done, so as firms navigate this new territory, success in the rapidly changing digital landscape will depend on their ability to carefully consider and adapt.

The tourism business is undergoing a dramatic transition (Arora et al., 2023a). The emergence of the metaverse brings novel approaches to discovering, experi-encing, and promoting tourist destinations. Through the immersive environment offered by this virtual world, people can explore locations, landmarks, and destina-tions without actually being there. The metaverse has the potential to revolutionize the way consumers research, book, and interact with travel experiences, which is good news for travel agencies. Offering immersive trip previews and experiences is one of the metaverse's most important effects on the travel industry. Potential travelers can now explore places from the comfort of their homes thanks to tech-nologies that imitate travel experiences, such as VR and AR. This gives a taste of what a place has to offer, affecting judgments about where to go and possibly drawing more tourists. In addition, tourism companies can build virtual versions of historical landmarks or real-world destinations thanks to the metaverse. This makes it possible to preserve and present cultural material in a way that is both very immersive and approachable. It can draw people who otherwise might not have had the chance to physically visit these destinations, fostering a greater interest in and appreciation for many cultures and locales.

Furthermore, the idea of tour guides and travel agencies is redefined by the metaverse. These companies may use virtual spaces to curate and market travel experiences in a way that is more dynamic and interesting. They can design immer-sive experiences that showcase hotels, events, and destinations in place of conven-tional brochures or webpages. Potential travelers' decision-making process may be greatly impacted by this immersive and customized marketing strategy. Moreover, "experiential travel" can be redefined in the metaverse. Businesses can use the virtual world to create engaging, instructive, and amusing experiences rather than just providing a glimpse of their destinations. Before making travel arrangements, tourists can acquire a taste of the real thing by immersing themselves in local cus-toms, taking part in activities, or even attending virtual events.

Changing Consumer Experiences and Engagement

The metaverse is redefining customer engagement, which is one of the biggest effects on businesses. Businesses are experimenting with new approaches to break beyond conventional limits and engage their audience through immersive experiences. The metaverse has given businesses a new platform for customer engagement, ranging from gamified marketing methods and experiential content in entertainment to interactive virtual shops in retail. The metaverse radically changes how people interact with and view vacation locations, having a profound impact on customer experiences and involvement within the tourist sector. First and foremost, the metaverse offers incredibly dynamic and immersive virtual worlds that change consumer experiences. Without actually going to a place, tourists can explore it in unprecedented detail and experience what it's like to be there. Through this immersive experience, visitors can gain a deeper understanding of a location, enabling them to make more educated travel arrangements. It transforms engaging with brochures or websites passively into actively exploring places with others. Furthermore, in facilitating personalized and interactive experiences, the metaverse encourages a new degree of customer involvement. Businesses in the tourism industry are able to provide customized experiences based on individual preferences because of the use of augmented reality and virtual reality technologies (Koo et al., 2023). For example, visitors can personalize their virtual tours by choosing particular sites or activities to experience. Customers feel more engaged and involved in the planning and discovery process when they receive this level of customization, which increases engagement. Furthermore, the metaverse promotes experiential travel, which redefines how tourists interact with locales. Travelers can virtually engage in local customs, cultural events, and activities instead of just seeing or reading about them. A stronger connection and comprehension of the place are fostered by this hands-on involvement, which offers a taste of the real experience. It thereby affects how customers behave and make decisions, which may increase their desire to visit the actual physical venues. Moreover, the metaverse facilitates an ongoing and easily accessible interaction with locations. Without any physical restrictions, users can return to virtual locations several times, interact with updated material, or take advantage of fresh opportunities. Travelers and locations develop a relationship over time as a result of this constant interaction, which sustains interest and may encourage more travel in the future. The authenticity and sensory pleasures of real travel are difficult to replicate in the virtual world, though. Even while the metaverse provides immersive experiences, it might not be able to adequately convey the tastes, smells, and overall "feel" of a place. This discrepancy may have an impact on consumers' perceptions as well as their actual travel experiences. The metaverse has a significant impact on how consumers engage and experience the tourism business. It improves visitor interactions with sites by providing immersive, customized, and engaging experiences. Maintaining consumer interest and advancing the tourist industry's future will depend on striking a balance between virtual and real-world travel experiences as the metaverse develops.

The Development of Trade and Income

The distinctions between the physical and digital worlds are becoming hazier due to the metaverse, which is changing the face of trade. Businesses are investigating digital asset ownership, developing new marketplaces, and investing in virtual real estate. Businesses are starting to offer both tangible and intangible commodities through e-commerce in the metaverse, which is making it necessary to reevaluate conventional income streams and business models.

Creative Tasks and Teamwork

There is a shift taking place in the way businesses function, particularly with regard to work and teamwork. The metaverse provides options for co-working spaces, virtual meetings, and remote work, enabling more immersive and engaging teamwork environments. It throws doubt on the conventional office layout and provides flexibility in the way and location of work.

Instruction, Practice, and Modeling

Businesses and educational institutions are embracing the metaverse because of its promise to offer realistic simulations and immersive learning environments. Knowledge transfer and skill development are being redefined by the metaverse, which offers everything from training simulations for different industries to virtual classrooms.

Originality/Value

Very few studies are available in the area of "Metaverse and Tourism" as the concept is new and less literature is available in this field.

This will be a novel piece of research, highlighting the areas of present research.

Limits

The biggest limitation of this work is that it is considering only one database, that is, SCOPUS and the rest of the information is opinion-based.

Purpose, Design, and Methodology

This study will attempt to find out what holds in the research on "metaverse and tourism" by reviewing the existing literature available in the SCOPUS database.

It will also explore how the reported content highlights the use of metaverse in different fields/dimensions of tourism. The study also attempts to analyze the bibliometric data extracted from SCOPUS to conduct, partial performance analysis, thematic analysis, and content analysis of present literature on "metaverse and tourism."

In the initial search process, the term "metaverse" AND "Tourism" OR "Virtual World in Tourism" OR "Virtual reality" AND "Tourism" has been used in the "source title" of the SCOPUS database, as of November 7, 2023. In the refinement process, before filtration, the results came out in the database between 2015 and

2023 with a total of 98 documents. After extracting articles and reviews from journals published in English language only, finally, 57 documents were retrieved. The authors have performed abstract and keyword analysis. Manual content analysis has been performed to build the base and future orientations in this field. The visual representations of the data have been performed by using either Vosviewer or R studio software.

Results

Table 7.1 shows the documents from all the fields retrieved from SCOPUS and an interesting fact is that, out of the list of 57 documents more than 95% of documents highlighted in the following are from the tourism and hospitality sector, highlighting the increasing interest of the authors in the field of metaverse in relation to the tourism industry. The main themes of the individual paper have also been highlighted.

While the keyword analysis was done for the dataset retrieved from the SCOPUS database, the major themes highlighted by the research community which emerged in relation to metaverse are education, healthcare, hospitality, tourism, media, smart tourism, sustainable tourism, virtual reality, blockchain, technology adoption, heritage tourism, virtual reality, virtual tourism, metaverse tourism, tourism development, tourism marketing, and cultural heritage. This fact is supported by Table 7.1 also, where the researchers have shown their interest in topics such as virtual tourism and cultural and heritage tourism in relation to the world of metaverse. If we look at the country's scientific production, the highest number of publications in the field of metaverse (Table 7.2) are done by China, followed by Malaysia. At rank third and fourth are South Korea and USA with 14 publications each. India, Spain, and Turkey have each published 11 articles, up to the date of data retrieved. UK and Thailand are ranked eighth and ninth, and Italy is ranked tenth. If we look at the production of universities, UCSI university is at the top, followed by the University of Macau. Swansea University is ranked third with five articles, along with rank fourth by the University of Central Florida. After that figures, AMERICAN UNIVERSITY OF SHARJAH, DOKUZ EYLÃœL UNIVERSITY, SHANDONG UNIVERSITY, SUAN DUSIT UNIVERSITY, THE HONG KONG POLYTECHNIC UNIVERSITY, and UNIVERSITAT POLITÃˆCNICA DE VALÃˆNCIA published four documents each on the date stated earlier.

The metaverse has the potential to completely change the way people travel, discover, and interact with destinations (Gursoy et al., 2022). It has the potential to revolutionize the tourist sector in a number of ways. The metaverse is redefining travel experiences, which is one of its main effects on the travel and tourism industry. With the aid of AR and VR technologies, people can become fully immersed in realistic, interactive representations of various locations. This gives prospective tourists a preview that goes beyond conventional brochures or webpages by enabling them to explore and experience locations they might be thinking about visiting. Travel decisions are influenced by this immersive experience, which may heighten interest in particular locations. In addition, the metaverse provides

Table 7.1 The total number of publications on Metaverse and Tourism in all disciplines out of the SCOPUS database as of November 7, 2023

	Authors	Year	Source title	Cited by	Themes/keywords
1.	Buhalis D.; Lin M.S.; Leung D.	2023	International Journal of Contemporary Hospitality Management	82	Augmented reality; Co-creation; Hospitality; Metaverse; Mixed reality; Virtual reality
2.	Rameshwar J.R.; King G.S.	2023	Journal of Metaverse	2	Caribbean; Extended Reality (XR); Industry 4.0 (I4.0); Metaverse; Survey
3.	Fazio G.; Fricano S.; Iannolino S.; Pirrone C.	2023	Information Technology and Tourism	0	Metaverse; Q-Methodology; Stakeholders; Tourism
4.	Jafar R.M.S.; Ahmad W.	2023	Tourism Review	0	Cognitive processing; Enjoyment; Escapism; Immersion; Metaverse tourism; PLS-SEM
5.	Kou G.; Yüksel S.; Dinçer H.	2023	Applied Soft Computing	0	Collaborative filtering; Metaverse; Neuro decision-making; Quantum spherical fuzzy sets; Sustainable investments
6.	Li J.; Hu P.; Cui W.; Huang T.; Cheng S.	2023	Information Technology and Tourism	0	Marine reconstruction; Meta-ocean; Metaverse; Virtual reality
7.	Lee U.-K.	2022	Sustainability (Switzerland)	22	information system success model; media richness theory; metaverse; smart tourism; virtual reality
8.	Garrido-Iñigo P.; Rodríguez-Moreno F.	2015	Interactive Learning Environments	29	effectiveness; French for specific purposes; self-identity; time spent/reuse ratio; virtual worlds
9.	Wei W.	2023	Journal of Hospitality and Tourism Insights	1	Hospitality; Metaverse; Second life; Tourism; Virtual reality
10.	Wong L.-W.; Tan G.W.-H.; Ooi K.-B.; Dwivedi Y.K.	2023	International Journal of Contemporary Hospitality Management	0	Critical reflection; Hospitality and tourism; Metaverse
11.	Behera R.K.; Bala P.K.; Rana N.P.	2023	Information Technology and Tourism	0	Metaverse; Metaverse tourism opportunities; Perceived barriers; Perceived benefits; Technology adoption; Tourism 4.0

(Continued)

Table 7.1 (Continued)

	Authors	Year	Source title	Cited by	Themes/keywords
12.	Huang X.-T.; Wang J.; Wang Z.; Wang L.; Cheng C.	2023	Frontiers in Psychology	0	experimental research; temperature comfort; thermal sensation; tourism spatial situation; virtual tourism
13.	Pellegrino A.; Wang R.; Stasi A.	2023	Heliyon	1	Consumption; Metaverse; Review
14.	Zhu C.; Wu D.C.W.; Hall C.M.; Fong L.H.N.; Kou-paei S.N.; Lin F.	2023	International Journal of Tourism Research	4	mental imagery; metaverse; non-immersive virtual reality; satisfaction; telepresence; travel intention; vividness
15.	Mladenović D.; Ismagilova E.; Filieri R.; Dwivedi Y.K.	2023	International Journal of Contemporary Hospitality Management	0	Credibility; Hospitality; Media richness theory; Metaverse; metaWOM; NTT; Persuasiveness; Reviewchain; Sensory; Tourism; WOM; Word-of-mouth
16.	Zhang X.; Yang D.; Yow C.H.; Huang L.; Wu X.; Huang X.; Guo J.; Zhou S.; Cai Y.	2022	Electronics (Switzerland)	16	cultural heritage; digital twin; five dimensions; metaverse; virtual reality continuum
17.	Çolakoğlu Ü.; Anış E.; Esen Ö.; Tuncay C.S.	2023	Journal of Hospitality and Tourism Insights	0	Metaverse; Tourists' VR experiences; VR applications in tourism
18.	Ioannidis S.; Kontis A.-P.	2023	Information Technology and Tourism	0	Avatar; Metaverse; NFT; Tourism; Travel; Trust
19.	Buhalis D.; O'Connor P.; Leung R.	2023	International Journal of Contemporary Hospitality Management	34	Hospitality ecosystem; Research directions; Smart hospitality
20.	Chu C.-H.	2023	Information Technology and Tourism	0	Blockchain; Deep learning; Massively multi-player online finance; Resource allocation; Tourism gamification
21.	Filimonau V.; Ashton M.; Stankov U.	2022	Journal of Tourism Futures	11	Consumer behavior; Consumer experience; Metaverse; Virtual reality; Virtual spaces

(Continued)

Table 7.1 (Continued)

	Authors	Year	Source title	Cited by	Themes/keywords
22.	Nam K.; Baker J.; Dutt C.S.	2023	Information Technology and Tourism	0	Hedonic; Heritage tourism; Metaverse; Partial least squares (PLS); Utilitarian; VR; Virtual tourism
23.	Zhang J.; Quoquab F.; Mohammad J.	2023	Tourism Review	0	Generation Y; Generation Z; Metaverse; Metaverse tourism; Self-determination theory; Theory of planned behavior
24.	Chen C.F.	2023	Critical Arts	0	AI; cultural practice; discourse carnival; literary theory; Metaverse
25.	Rejeb A.; Rejeb K.; Treiblmaier H.	2023	Information (Switzerland)	1	education; healthcare; immersive digital environment; metaverse; research agenda; research questions; tourism
26.	Su P.-Y.; Hsiao P.-W., Fan K.-K.	2023	Sustainability (Switzerland)	0	behavioral intention; learning effects; partial least squares multigroup analysis; sustainable education; virtual reality
27.	Wei D.	2022	International Journal of Geoheritage and Parks	25	blockchain; Gemiverse; metaverse; professional certification; tourism platform
28.	Luo Z.; Li Y.; Semmen J.; Rao Y.; Wu S.-T.	2023	Light: Science and Applications	0	
29.	Sánchez-Amboage E.; Enrique Membiela-Pollán M.; Martínez-Fernández V.-A.; Molinillo S.	2023	Museum Management and Curatorship	0	COVID-19; Europe; Meta; metaverse; museums; Tourism marketing
30.	Fan Z.; Chen C.; Huang H.	2022	Heritage Science	6	Cultural heritage; Digital documentation; Metaverse; Tourism service; Virtual reality; Zhu Xi
31.	Buhalis D.; Leung D.; Lin M.	2023	Tourism Management	71	Immersive experience; Information communication technologies; Metaverse; Virtual experience

(Continued)

Table 7.1 (Continued)

	Authors	Year	Source title	Cited by	Themes/keywords
32.	Gursoy D.; Malodia S.; Dhir A.	2022	Journal of Hospitality Marketing and Management	116	Internet 3.0; metaverse; metaverse tourism; virtual tourism
33.	Koohang A.; Nord J.H.; Ooi K.-B.; Tan G.W.-H.; Al-Emran M.; Aw E.C.-X.; Baabdullah A.M.; Buhalis D.; Cham T.-H.; Dennis C.; Dutot V.; Dwivedi Y.K.; Hughes L.; Mogaji E.; Pandey N.; Phau I.; Raman R.; Sharma A.; Sigala M.; Ueno A.; Wong L.-W.	2023	Journal of Computer Information Systems	58	banking services; education; healthcare; hospitality; human resource; management; manufacturing; marketing; Metaverse; operations management; the retailing industry; tourism
34.	Ud Din I.; Almogren A.	2023	Information Technology and Tourism	0	Disability; Mental health; Metaverse; Virtual reality; Well-being
35.	Go H.; Kang M.	2023	Tourism Review	23	Generation Z and alpha; Google trends; Metaverse tourism; Sustainable tourism; UNWTO SDGs; Virtual reality
36.	Ariza-Montes A.; Quan W.; Radic A.; Yu J.; Han H.	2023	Journal of Travel and Tourism Marketing	0	behavioral models; conference/meeting industry; human values; Metaverse; portrait values questionnaire; structural equation model (SEM); technology innovation; the unified theory of acceptance and use of technology (UTAUT2) model; tourism industry; traveler behaviors
37.	Özdemir Uçgun G.; Şahin S.Z.	2023	Current Issues in Tourism	0	Metaverse; sustainability; technology; tourism; tourism development; virtual reality

(Continued)

Table 7.1 (Continued)

	Authors	Year	Source title	Cited by	Themes/keywords
38.	Baker J.; Nam K.; Dutt C.S.	2023	Information Technology and Tourism	0	Heritage tourism; Interview; Metaverse; User experience; VR
39.	Zaman U.; Koo I.; Abbasi S.; Raza S.H.; Qureshi M.G.	2022	Sustainability (Switzerland)	24	COVID-19 travel anxiety; digital tourism sustainability; metaverse; post-COVID-19 tourism; space travel; tech savviness; travel FOMO; virtual reality
40.	Yang F.X.; Wang Y.	2023	Journal of Hospitality and Tourism Research	6	metaverse; metaverse tourism; synthetic reality; taxonomy; virtual world
41.	Gursoy I.T.; Aktas E.; Tecim V.; Kurgun O.A.	2023	Information Technology and Tourism	0	Event management; Hospitality; Metaverse; System development life cycle
42.	Kang H.-C.; Baek W.-Y.; Choi J.-Y.; Kim J.-S.	2023	Journal of Marine and Island Cultures	0	digital transformation; island tourism; revitalization
43.	Ampountolas A.; Menconi G.; Shaw G.	2023	Tourism Economics	4	blockchain; hospitality; Metaverse; online travel agencies; tourism; virtual reality
44.	Chen S.; Chan I.C.C.; Xu S.; Law R.; Zhang M.	2023	Journal of Travel and Tourism Marketing	3	Asia Pacific; destination competitiveness; drivers; grounded theory; hindrances; Metaverse tourism; qualitative methodology; stakeholder theory; tourism development; tourism marketing
45.	Lin K.J.; Ye H.; Law R.	2023	Information Technology and Tourism	1	Blockchain; Institutional theory; Metaverse; Stakeholder; Technology adoption; Tourism destination
46.	Jo H.	2023	Information Technology and Tourism	1	Continuance intention; Hedonic benefits; Metaverse; Symbolic benefits; Utilitarian benefits

(Continued)

Table 7.1 (Continued)

	Authors	Year	Source title	Cited by	Themes/keywords
47.	Bilgihan A.; Ricci P.	2023	Journal of Hospitality and Tourism Technology	0	Artificial intelligence (AI) chatbots; Emerging technologies; Hospitality industry; Hotel sales and marketing; Metaverse; Revenue optimization; Virtual reality
48.	Suanpang P.; Niamsorn C.; Pothipassa P.; Chunhapataragul T.; Netwong T.; Jermsittiparsert K.	2022	Sustainability (Switzerland)	13	extensible metaverse; smart tourism; tourism intention; virtual reality
49.	Zhang J.; Quoquab F.	2023	International Journal of Tourism Cities	1	Chinese traditional culture; Games; Metaverse tourism; Urban tourism
50.	Dutta D.; Srivastava Y.; Singh E.	2023	Information Technology and Tourism	1	Human–machine agency; Learning & development; Metaverse; Practice theory
51.	Monaco S.; Sacchi G.	2023	Sustainability (Switzerland)	10	digital innovation; food; Metaverse; phygital tourism; research; tourism; Web 3.0; wine
52.	Montoya Esquer J.E.; Lara López G.	2023	Virtual Reality	0	Natural language processing; Qwerty paradigm; Text entry; User input; Virtual reality
53.	Corne A.; Massot V.; Merasli S.	2023	Information Technology and Tourism	0	Blockchain; fsQCA method; Technology Acceptance Model (TAM); Tourism sector
54.	Tsai S.-P.	2022	Journal of Vacation Marketing	16	actual visit intention; holistic happiness; holistic presence; Metaverse marketing; metaverse tours
55.	Marti-Testón A.; Muñoz A.; Gracia L.; Solanes J.E.	2023	Applied Sciences (Switzerland)	1	city; guided tours; spatial; touristic promotion; virtual exhibitions
56.	Zhang W.; Wang Y.	2023	Information Technology and Tourism	0	Destination image; Metaverse; Storytelling; Visit intention
57.	Hui X.; Raza S.H.; Khan S.W.; Zaman U.; Ogadimma E.C.	2023	Sustainability (Switzerland)	2	dispositional empathy; eco-literacy; immersive journalism; media richness theory; metaverse; pro-environmental theory; regenerative tourism; sustainable tourism

Source: Created by authors from the SCOPUS database

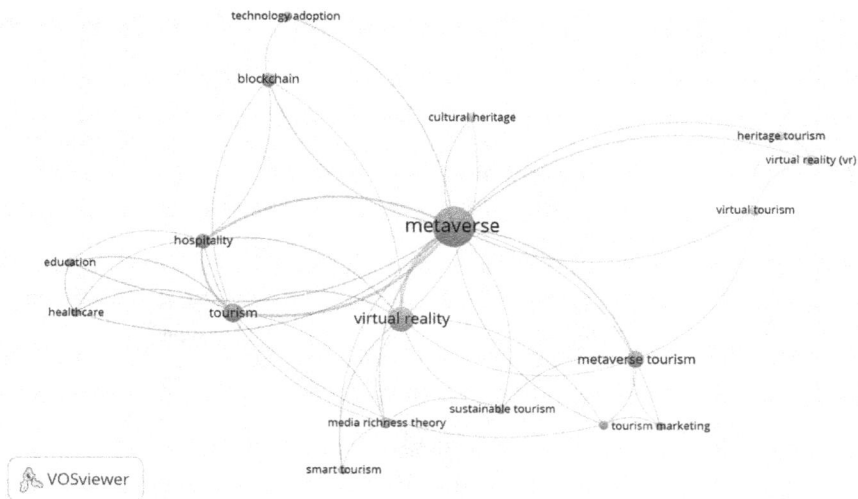

Figure 7.1 The keyword analysis.

Table 7.2 Countries' scientific production and highest production of universities

Rank	Region	Freq	Rank	Affiliation	Articles
Rank 1	CHINA	29	Rank 1	UCSI UNIVERSITY	8
Rank 2	MALAYSIA	15	Rank 2	UNIVERSITY OF MACAU	8
Rank 3	SOUTH KOREA	14	Rank 3	SWANSEA UNIVERSITY	5
Rank 4	USA	14	Rank 4	UNIVERSITY OF CENTRAL FLORIDA	5
Rank 5	INDIA	11	Rank 5	AMERICAN UNIVERSITY OF SHARJAH	4
Rank 6	SPAIN	11	Rank 6	DOKUZ EYLÃœL UNIVERSITY	4
Rank 7	TURKEY	11	Rank 7	SHANDONG UNIVERSITY	4
Rank 8	UK	10	Rank 8	SUAN DUSIT UNIVERSITY	4
Rank 9	THAILAND	8	Rank 9	THE HONG KONG POLYTECHNIC UNIVERSITY	4
Rank 10	ITALY	7	Rank 10	UNIVERSITAT POLITÃ^CNICA DE VALÃ^NCIA	4

Source: Created by authors by using the SCOPUS database

a special venue for advertising and marketing tourism spots. Companies in the tourist sector have the ability to produce and present immersive virtual experiences, which may be used to market lodging, activities, and attractions in a more dynamic and interesting way. This can have a significant impact on prospective travelers' decisions and how they organize their travels. Moreover, the metaverse possesses the capability to transform the notion of destination development and planning (Um et al., 2022). The metaverse can be used by developers and urban planners to envision, build, and test new urban areas, resorts, and infrastructure. This can facilitate more effective and sustainable development by improving our understanding of the effects on the environment and society prior to actual implementation. Nonetheless, there are still issues, especially with regard to standardization and accessibility of technology. It is vital to guarantee that the technology required for a flawless virtual tourist experience is widely accessible. Furthermore, maintaining the quality and security of these experiences as well as standardizing them across various platforms are crucial factors for the effective integration of the metaverse in the travel and tourism sector. The tourist sector is greatly impacted by the metaverse, which provides new avenues for destination exploration, marketing, and experience (Wei, 2024). Technology will have a significant impact on how travel and tourism develop in the future, posing both opportunities and challenges for both businesses and tourists as the metaverse and travel technologies develop. Due to COVID-19, distress not only caused challenges for the tourism sector but has opened up new avenues also as people realized the importance of spiritual tourism, well-being, and travel in many respects (Arora et al., 2021, 2023b).

The metaverse's influence on the travel and tourism sector has been more like a slow-moving revolution than a big upheaval in any one nation. Although the metaverse has demonstrated the ability to change how people research travel locations and how companies advertise their services, its significant influence on the tourism sector of any given nation did not seem immediately apparent at the time in question. A number of nations looked into the incorporation of virtual reality and other technology into their tourism marketing efforts as the idea of the metaverse gained traction. However, as of yet, neither the metaverse's widespread adoption nor its effects have been significant enough to identify any one nation over another. The USA, Japan, South Korea, and several European countries, which place a high priority on technical innovation and tourism, were among the first to adopt and experiment with immersive technologies to improve their tourism offerings. It nonetheless takes quite a while to see the extent to which the metaverse would change a nation's tourism sector from the ground up. The impact of the metaverse on international travel may change over time, and by now, some nations may have advanced significantly in using these technologies to improve traveler experiences. Up to this point, no clear evidence that the metaverse had a singular and significant impact on tourism in any specific nation is available. Out of the significant work published by several countries, the most cited countries come out to be UK, Norway, Malaysia, Korea, Spain, China, USA, Singapore, Thailand, and Monaco as shown in Table 7.3, as per the data extracted from SCOPUS.

Table 7.3 Most Cited Countries

Country	TC	Average article citations
UK	198	39.6
NORWAY	116	116
MALAYSIA	63	10.5
KOREA	49	7
SPAIN	30	10
CHINA	25	3.57
USA	24	6
SINGAPORE	16	16
THAILAND	14	7
MONACO	10	10

Source: Created by authors by using the SCOPUS database

Future Research Agenda

Future research projects have a bright future ahead of them attributable to the metaverse's development and integration with the tourism sector. Scholars may investigate a range of facets that may considerably augment and progress the metaverse's function inside the travel and tourist industry. Examining user behavior and experiences in the metaverse within the context of tourism is a crucial subject for future research. Crucial insights can be gained by comprehending how people connect, engage, and navigate virtual travel experiences. Investigating user behavior in virtual worlds from a psychological and sociological perspective may reveal trends, preferences, and the elements influencing travel destination decision-making. These studies can provide businesses with important insights for creating more successful and interesting virtual tourist experiences. In addition, there is much to learn about how the metaverse affects destination development and sustainable tourism strategies (Go & Kang, 2023). Scholars can look at how eco-friendly, sustainable travel destinations can be planned and designed using the metaverse. This could entail researching how the metaverse can be used to test and simulate sustainable tourist efforts, urban planning, and infrastructure, thereby promoting the development of tourism that is more environmentally sensitive. Furthermore, studies examining the metaverse's financial effects on the travel and tourism sector are crucial. It is imperative that governments and entrepreneurs alike comprehend the financial aspects of the virtual tourist industry, including revenue models and the possibility of new revenue streams. Such studies could examine the scalability of different virtual experiences, the economic feasibility of virtual tourism, and its effect on traditional tourism earnings. Furthermore, it is imperative to investigate the ethical and social ramifications of the metaverse in tourism. Scholars have the ability to investigate issues related to security, privacy, and ethics in relation to virtual experiences. Building trust and promoting the wider adoption of virtual tourism experiences requires an understanding of how to develop secure and reliable virtual environments while protecting user privacy

(Far & Rad, 2022; Hazy, 2012; Verma & Rao, 2016; Narayanswamy, 2016). Future studies on the metaverse in the context of tourism should look into user behavior, technological developments, sustainability, the consequences for the economy, and ethical issues. Future virtual tourist experiences will be greatly influenced by these research objectives, opening up more interesting, genuine, and sustainable travel options inside the metaverse.

Future Impacts of Metaverse

1. Entertainment and gaming will probably be the areas where the metaverse has the biggest immediate effects. It has the potential to completely transform gaming by providing more engaging and dynamic gameplay. Businesses in this industry could have to change to produce experiences and materials that are appropriate for the metaverse.
2. Because of virtual experiences and stores, the metaverse has the potential to transform the way people shop. To sell virtual or actual goods, businesses might create a presence in the metaverse, obfuscating the distinction between digital and physical trade.
3. Within the metaverse, virtual real estate is already starting to become in demand. Companies may spend money on virtual properties for events, advertising, or special experiences.
4. The metaverse has the potential to completely change how people work by enabling more engaging and dynamic distant collaboration. Companies can use the metaverse for simulations, co-working spaces, and virtual meetings.
5. Immersion learning opportunities and training simulations are provided by the metaverse. Businesses and educational institutions could use this technology for workshops, skill development, and training.
6. The metaverse could be used in the healthcare industry for remote consultations in a more immersive setting, professional training, and simulations.
7. For businesses to attract audiences in the metaverse, their marketing techniques may need to be modified. Product placements and interactive advertising in virtual environments may become increasingly widespread.
8. It will take cutting-edge infrastructure and technology to construct the metaverse. Companies that specialize in developing metaverse-related tools, platforms, and solutions will be important.
9. Legal and legislative frameworks will need to change as the metaverse develops to handle concerns about digital ownership, intellectual property rights, user privacy, and security.

In conclusion, the metaverse has the potential to have a big impact on a lot of different industries, opening up new doors for customer interaction, business model development, and innovation. Businesses in a variety of sectors will need to modify and advance their plans in order to take advantage of the potential this new digital arena offers.

Challenges and Opportunities

The metaverse does, however, come with some difficulties (Tang et al., 2022). Businesses need to handle critical issues such as digital ownership, privacy, security, and legal frameworks. Businesses have a variety of issues, including developing experiences and material appropriate for the metaverse, making sure that users have a flawless user experience, and keeping up with the rapidly changing technology landscape. New dimensions for customer involvement, commerce, work, education, and other areas are being made possible by the metaverse, which is changing the corporate landscape. It is a change in how companies see and produce value, not just a technical innovation. Adoption of this digital frontier necessitates creativity, adaptability, and a thorough comprehension of the potential and difficulties it brings. Enterprises that recognize and capitalize on the metaverse's promise are well-positioned to spearhead the subsequent wave of revolutionary transformation.

Discussion and Concluding Remarks

Although there are still obstacles to overcome, the metaverse has the potential to significantly affect many other businesses, including tourism. The quality of the virtual experiences, adoption rate, and accessibility of the technology are critical factors in the metaverse's potential in the tourism sector. To guarantee a smooth and interesting virtual tourism experience, bandwidth, technological constraints, and the requirement for high-quality content generation are essential elements that must be addressed. The metaverse has a revolutionary effect on the travel and tourism industry, opening up new channels for destination marketing, traveler experience, and travel destination sales. It has the power to completely transform how people travel and connect with the world, ushering in a new era of highly immersive, interactive, and widely available tourism. The tourist sector will probably see a dramatic change as the metaverse develops more, changing how we travel and see the world. The transformative influence of the metaverse on tourism and related industries surpasses just virtual discovery and advertising. It radically transforms the way companies in this industry function and provide services, which has the effect of completely changing the traveler experience as a whole. The possibility for improved consumer feedback and engagement is one important factor. Businesses can learn a great deal about the tastes and behavior of their customers by leveraging the metaverse. Virtual environment interaction analysis can yield priceless information that helps companies better customize their products to the preferences of their target market. This can therefore result in more individualized and client-focused travel services. The field of destination development and planning is another important dimension. Urban planners, architects, and developers have a cutting-edge platform to design and test new tourist destinations because of the metaverse. Infrastructure, resorts, attractions, and city planning may all be seen and improved in a virtual environment. This can help with more efficient and sustainable development that takes into account social and environmental effects before being implemented in the real world. In addition, the metaverse provides a way for the tourism sector to diversify its sources of income. In addition to conventional methods like reservations and in-person visits, businesses can also make money through online events, virtual experiences, and unique digital content. This can increase the accessibility and inclusion of the tourism industry by appealing to a larger audience, including those who might not be able to visit physically. The metaverse also promotes alliances and cooperation across industry players. It offers a venue for many organizations to work together to create immersive experiences, such as hotels, airlines, tour operators, and neighborhood businesses. By working together, places may become more appealing overall and provide travelers with a more seamless and all-encompassing vacation experience. The metaverse may also have an impact on hiring and training practices in the travel and tourism sector. The need for people with experience in developing and overseeing virtual experiences is increasing as the industry develops in tandem with technology breakthroughs. This calls for the creation of new positions and educational opportunities for experts, such as virtual reality experience designers, metaverse-focused digital marketers, and virtual tourism customer experience managers. However, problems still exist. Adoption of metaverse technology necessitates large investments in software development, infrastructure, and content production. Furthermore, to guarantee a secure and reliable user experience, problems with data security, privacy issues, and standardizing experiences across various platforms must be resolved. In conclusion, the effects of the metaverse on travel and related industries extend beyond online research and advertising. It affects a variety of areas, including personnel development, collaboration, income sources, destination planning, and consumer engagement. Businesses must adjust to these shifts as they traverse this revolutionary terrain if they hope to fully realize the metaverse's potential to completely revolutionize the travel and tourism sector.

References

Arora, M. (2020). Post-truth and marketing communication in technological age. In *Handbook of research on innovations in technology and marketing for the connected consumer* (pp. 94–108). IGI Global.

Arora, M., Dhiman, V., & Sharma, R. L. (2023a). Exploring the dimensions of spirituality, wellness and value creation amidst Himalayan regions promoting entrepreneurship and sustainability. *Journal of Tourismology*, 9(2), 86–96. https://doi.org/10.26650/jot.2023.9.2.1327877

Arora, M., Kumar, J., & Valeri, M. (2023b). Crises and resilience in the age of digitalization: Perspectives of past, present and future for tourism industry. In *Tourism innovation in the digital era* (pp. 57–74). Emerald Publishing Limited.

Arora, M., & Sharma, R. L. (2021). Repurposing the role of entrepreneurs in the havoc of COVID-19. In *Entrepreneurship and big data* (pp. 229–250). CRC Press.

Arora, M., & Sharma, R. L. (2022a). Coalescing skills of gig players and fervor of entrepreneurial leaders to provide resilience strategies during global economic crises. In *COVID-19's Impact on the cryptocurrency market and the digital economy* (pp. 118–140). IGI Global.

Arora, M., & Sharma, R. L. (2022b). Integrating gig economy and social media platforms as a business strategy in the era of digitalization. In *Integrated business models in the digital age: Principles and practices of technology empowered strategies* (pp. 67–86). Springer International Publishing.

Arora, M., & Sharma, R. L. (2023). Artificial intelligence and big data: Ontological and communicative perspectives in multi-sectoral scenarios of modern businesses. *Foresight, 25*(1), 126–143.

Arora, M., Sharma, R. L., & Walia, S. (2021). Revisiting the inner self in times of debilitating distress: Gateways for wellness through spiritual tourism. *International Journal of Religious Tourism and Pilgrimage, 9*(5), 4.

Buhalis, D., Leung, D., & Lin, M. (2023). Metaverse as a disruptive technology revolutionising tourism management and marketing. *Tourism Management, 97*, 104724.

Far, S. B., & Rad, A. I. (2022). Applying digital twins in metaverse: User interface, security and privacy challenges. *Journal of Metaverse, 2*(1), 8–15.

Go, H., & Kang, M. (2023). Metaverse tourism for sustainable tourism development: Tourism agenda 2030. *Tourism Review, 78*(2), 381–394.

Gursoy, D., Malodia, S., & Dhir, A. (2022). The metaverse in the hospitality and tourism industry: An overview of current trends and future research directions. *Journal of Hospitality Marketing & Management, 31*(5), 527–534.

Hazy, J. K. (2012). Leading large: Emergent learning and adaptation in complex social networks. *International Journal of Complexity in Leadership and Management, 2*(1/2), 52–73. https://doi.org/10.1504/IJCLM.2012.050395

Kim, J. (2021). Advertising in the metaverse: Research agenda. *Journal of Interactive Advertising, 21*(3), 141–144.

Koo, C., Kwon, J., Chung, N., & Kim, J. (2023). Metaverse tourism: Conceptual framework and research propositions. *Current Issues in Tourism, 26*(20), 3268–3274.

Kye, B., Han, N., Kim, E., Park, Y., & Jo, S. (2021). Educational applications of metaverse: Possibilities and limitations. *Journal of Educational Evaluation for Health Professions,* 18.

Lee, L. H., Braud, T., Zhou, P., Wang, L., Xu, D., Lin, Z., Kumar, A., Bermejo, C., & Hui, P. (2021). All one needs to know about metaverse: A complete survey on technological singularity, virtual ecosystem, and research agenda. *arXiv preprint,* arXiv:2110.05352.

Narayanswamy, R. (2016). Leadership is not a destination but a place to come from Gandhi's contribution to evolutionary excellence. *International Journal of Complexity in Leadership and Management, 3*(4), 278–283. https://doi.org/10.1504/IJCLM.2016.087151

Özdemir Uçgun, G., & Şahin, S. Z. (2023). How does metaverse affect the tourism industry? Current practices and future forecasts. *Current Issues in Tourism,* 1–15.

Park, S. M., & Kim, Y. G. (2022). A metaverse: Taxonomy, components, applications, and open challenges. *IEEE Access, 10*, 4209–4251.

Rather, R. A. (2023). Metaverse marketing and consumer research: Theoretical framework and future research agenda in tourism and hospitality industry. *Tourism Recreation Research,* 1–9.

Tang, F., Chen, X., Zhao, M., & Kato, N. (2022). *The roadmap of communication and networking in 6G for the metaverse.* IEEE Wireless Communications.

Um, T., Kim, H., Kim, H., Lee, J., Koo, C., & Chung, N. (2022, January). Travel Incheon as a metaverse: Smart tourism cities development case in Korea. In *ENTER22 e-Tourism conference* (pp. 226–231). Springer International Publishing.

Valeri, M. (2023). *Tourism innovation in the digital era. Big data, AI and technological transformation.* Emerald Publishing, UK.

Valeri, M. (2024). *Innovation strategies and organizational culture in tourism. Concepts and case studies on knowledge sharing.* Routledge Publishing.

Verma, P., & Rao, M. K. (2016). Authentic leadership approach for enhancing innovation capability: A theoretical investigation. *International Journal of Complexity in Leadership and Management, 3*(4), 284–300. https://doi.org/10.1504/IJCLM.2016.087114

Wang, Y., Su, Z., Zhang, N., Xing, R., Liu, D., Luan, T. H., & Shen, X. (2022). A survey on metaverse: Fundamentals, security, and privacy. *IEEE Communications Surveys & Tutorials, 25*(1), 319–352.

Wei, W. (2024). A buzzword, a phase or the next chapter for the internet? The status and possibilities of the metaverse for tourism. *Journal of Hospitality and Tourism Insights, 7*(1), 602–625.

Part II
Experiences in Metaverse

8 Tourism Agenda 2030 in Greece

Evangelia Kasimati, Athina Rentifi,
and Nikolaos Vagionis

Introduction

The tourism sector significantly contributes to Greece's gross domestic product (GDP), serving as a vital source of income, both directly and indirectly, while also engaging with various economic sectors. However, in the modern era, the tourism product cannot remain static. New factors, such as climate change and environmental protection, pose risks, while digitalization and technological progress offer opportunities to enhance the tourism experience. Traditionally, Greek tourism has revolved around leisure, particularly the "sun and beach" concept, which primarily focused on the islands and overlooked the unique characteristics of different regions, such as gastronomy, thermal waters, mountain activities, sports tourism, and agrotourism, which could offer exceptional experiences. This research explores Greece's tourism agenda for 2030, aiming to increase visitor numbers, receipts, and average spending per trip, and enhance the overall quality of the tourism product and experience.

This chapter examines Greek tourism toward 2030. The megatrends that already shape the changes in the global and – consequently – in Greek tourism are the main topic of the study. Such megatrends are digital upgrading and transformation, environmental protection and sustainability, social and demographic changes, overtourism issues, sharing economy, emerging and secondary destinations, security and crisis management, as well as the COVID pandemic and future health issues.

Digital technology has become integral to the tourism industry, with many businesses embracing modern technology to enhance their competitiveness. Destinations are creating mechanisms to provide expertise and resources to these businesses, enabling them to adapt to digital transformation. Sustainability, encompassing environmental, economic, social, and cultural dimensions, has become a key component of national tourism policies and business practices. Destinations are implementing initiatives to promote sustainability and seeking international certifications to demonstrate their commitment to these principles. Demographic shifts, including population growth and aging, as well as evolving preferences and trends among different age groups, are influencing global tourism. Traveler preferences are changing, and the middle class's rising disposable income is driving

DOI: 10.4324/9781003497004-10

demand for personalized services, necessitating adaptation in the tourism sector to cater to individual preferences.

On the other side, long-term predictions indicate an "explosion" in travel, potentially causing congestion in popular destinations and adverse impacts on the environment, society, and culture. Solutions to address overtourism include promoting less burdened or emerging destinations, implementing special taxes and fines to mitigate the phenomenon, and regulating traffic while offering accommodations and activities in popular areas. Additionally, the rise of the sharing economy requires regulation to ensure fair competition and safety for citizens and tourists alike.

Emerging destinations, often less frequented by tourists, are expected to gain popularity as travelers seek unique experiences. These destinations are developing specialized tourism strategies to position themselves effectively in the global market. Negative events and crises, such as natural disasters, terrorist attacks, epidemics, political instability, and social unrest, significantly impact tourism demand and traveler behavior. On the other side, the COVID-19 pandemic profoundly affected the tourism sector, emphasizing the importance of safety, protection, reliability, and personalized experiences, accelerating digital transformation and sustainability practices while leading to the emergence of new destinations.

On the basis of these megatrends, this study provides useful data and presents action plans to further develop Greek tourist products, in terms of destination and clusters across the country. After the introduction, eight sections are followed to present and analyze each of the aforementioned megatrends that play an important role in Greek tourism toward 2030. The study's methodological approach is based on a qualitative analysis, taking into account secondary data and research. Finally, the conclusions, challenges, and policy implications are presented.

Digital Upgrading and Transformation

Within the metaverse concept, digital transformation becomes an integrated part of the business ecosystem as it increases its competitiveness. Companies that are ahead with their digitalization increase not only their competitiveness but also their visibility from possible customers.

Digitalization upgrade includes the private as well as the public sector. Public support and funding for the development of digital applications are important. Funding from EU programs gave a big push to a portion of companies to accelerate their digital upgrading. The structure of the Greek economy is such that there are many small family enterprises. The least improvement has been to create a webpage or a profile in social networks or tourism-related applications. Apart from the support in tourism businesses to increase the use of digital tools, also the development of a single platform of suppliers of products and services benefits consumers. Specifically, tourism public authorities provide information through digital channels which makes it more accessible and faster. Along with the wide plan for the digitalization of the public sector, new services are added. In addition, the experience for each tourist is improving when bureaucracy hustles are processed online, or important information is easily accessible. Therefore, the satisfaction of the trip increases. Nevertheless, digitalization and technological progress offer new potentials that can advance the tourism experience. The main approach until recently for Greek tourism products focused on leisure, the so-called "sun and beach." However, that approach did not exhaust the potential of all geographical areas, was concentrated mainly on the islands, and did not highlight various particularities of each region.

Digital transformation can maximize customer satisfaction and increase collaboration and synergies across similar disciplines. When business collaboration accelerates, companies can enrich the offered product and the implementation of differentiated products. The increasing penetration of digital technologies results in increased satisfaction from the experience through ease in booking scheduling of travel and widening of choices, creation of online tourist portals, and their complete interconnection for all regions, tourist portals, and databases. Also, the adoption of new technologies minimizes direct physical contact with successful digital substitution through digital catalogs, reservations through applications, contactless payments, or biometric control that allows safe and hassle-free travel.

Nevertheless, positive is the development or expansion of digital services and digital marketing provided by public authorities and organizations in the communication channels with the consumer. The latest developments in technology changed the scenery that companies need to operate. Other areas that can change significantly the current product is the utilization of big data analytics and artificial intelligence (AI) for better market targeting and demand forecasting. The use of big data gives increased efficiency to this tool as advertisements are more targeted based on tourist profiles that have been created according to searches or purchases. Digital supervision can also decrease operational costs of businesses, for example, with the control of electricity or AC consumption – also, the promotion of the use of Internet of Things (IoT) and provision of free Wi-Fi in places with increased touristic interest and highlighting selected destinations as "smart." Already, Naxos Island has a plan to become the first smart island.

These aspects were diligently examined within the process of setting up a new fresh vision for Greek tourism with the ultimate goal of increasing visitors, receipts, and average spending per trip as well as qualitative advancement of the product and of the experience. The action plans that were constructed for each prefecture separately target to reveal its comparative advantage. Additionally, take into account the prerequisites needed in order to fulfill the objectives set which include the development and advancement of infrastructure, digital upgrade and transformation, enhancement of skills, and development of entrepreneurship.

Environmental Protection and Sustainability

In the current era, the design of tourism products cannot remain stagnant – on one hand, the risks that arise from climate change and, on the other hand, the need for environmental protection present new elements that need to be considered. Sustainable development includes a wide spectrum of factors. These refer to environmental, economic, social, and cultural aspects. Climate change accelerates the creation of policies for the protection of environment and sustainability. Also, new investment design is obliged to be environment friendly and compliant with the introduction of new rules and policies. It must be noted that customers would have been willing to pay more for products compliant with environmental protection.

Climate change is expected to have severe consequences for national economies with a GDP decrease, increased costs to handle these changes, and services and infrastructure being transformed. Temperature increase will increase the need and use of AC versus current use which requires more energy resources for accommodation and restaurants. Also, climate change can differentiate popular areas and tourist preferences. Actions for tourism season expansion will be assisted from climate change as the weather will improve for longer periods and it is possible that temperatures in current popular places to be unbearable during the current pick season. This way, there will be a redistribution among places, months, and prefectures. Already, tourists discover, apart from the islands, also mountainous areas.

Policies regarding tourism apart from the increased need for tourism services need to include actions to handle the global population increase and the availability-use of resources, availability of water, and protection from unexpected natural disasters. In this direction, tourism has to contribute to the optimal use of resources and the protection of national heritage and biodiversity. Also, principles have to be designed for the respect of cultural differences and tolerance among local communities and tourists. Also, environmental protection has to take into consideration the behavioral aspects of consumers, the increased risk of companies, the policies of societies that host tourists and possibly face crises, as well as the expected results of the extended use of technology.

Greece targets to increase by 2030 tourist arrivals. However, it cannot be omitted that this target requires qualitative changes in order to build the necessary infrastructure in terms of accommodation, transportation and connectivity, energy needs, water needs, waste management, quality and quantity of food, primary health care, protection in cases of extraordinary physical phenomena, and so on. In this direction, the cyclical economy and the reuse of waste are important for the best possible use of resources. The total cost of the implementation of necessary plans can reach 510 billion euros up to 2100 according to INSETE. Apart from the cost of protection, the cost also of reconstruction exists in areas severely hit by phenomena such as fires, earthquakes, floods, and so on for people, animals, public infrastructure, and private companies.

Nevertheless, tourism can enhance local income and increase employment in a variety of different fields. Popular areas rely on tourism income as it is the main annual source of income. Therefore, sustainability in the tourism sector is also related to contribution to local economies to develop further all sectors interrelated directly or indirectly to tourism. Many local communities already announced environmentally friendly policies, like, for example, the islands of Lipsi, Astypalea, and Tilos.

This framework led to the creation of international partnerships in tourism where tourism industry companies can develop common practices and training programs for employment in the sector. Policies toward environment protection and sustainability must be in line with 17 UN SDG policies as far as tourism is concerned.

Social and Demographic Changes

The target group of tourism entrepreneurs is not constant. Preferences change in line with demographic evolution or social differentiation, and tourism needs to adapt to different age or social groups.

Demographic changes include the aging population which is expected to continue to increase in the next years. Therefore, the needs of elderly people going on holiday compared with Millennials or Generation Z are different. Especially in the South of Europe, retirement villages are being created because of the weather. In accordance with that, the real estate market in Greece in the last years has been boosted by foreigners' purchases. According to projections, population aging will result in tailored services targeting this specific age group that will be in higher demand. These trends lead to the creation of the so-called silver economy because of the increase in life expectance as well as the improvement of the standards of living. Specifically, according to the United Nations, the age group over 60 is expected to increase to 2.1 billion people.

In addition, the needs and preferences of younger tourists versus the elderly are different. Until today, Greek tourism was based on the "sun and sea" product. The 2030 plan has included various other experiences like cultural or religious that cover third age's preferences. Nevertheless, younger people are more competent with the use of technology and digital services, and it is very possible to use travel websites or applications and to post their experiences in critics. Younger tourists are less likely to travel in groups-packages. Big data companies can have access to bulk information on Internet surfing or transactions that can assist them in designing new products in line with tourists' preferences. Also, digital marketing is being used by companies to approach new customers. Within this mindset are initiatives like the "Stay like a local" of Hyatt which can provide information less accessible.

Social changes also bring the need for changes. The pandemic, for example, changed completely hygiene rules. Also, the economic conditions that improve in emerging markets bring new target groups – potential customers including Asians and Africans where middle-income citizens increase. On the other hand, traditional markets where disposable income shrinks, because of inflation and energy costs, decrease their vacation abroad.

Moreover, another group of tourists who increase their presence and potential needs is digital nomads for whose special visas are issued. Digital nomads include employees with virtual offices who can work from anywhere in the world and combine business with local experience with the primary need for fast Internet. South of Europe, especially in summer but also in mild winter with decreased need for energy consumption, becomes more attractive for a longer stay. Already, companies from Northern Europe rent whole buildings for part of their employees in Greek islands like Crete.

Overall, all travelers seek for experience, authenticity, nature, culture, gastronomy, and tailor-made packages. The above developments were taken into consideration when designing the tourism 2030 agenda, and for each prefecture, different activities based on its landscape were designed. For example, hiking for different levels, underwater activities, gastronomy, or winning experiences. Also, social networks affect preferences and introduce new or less mass choices.

Overtourism

According to Koens et al. (2018), overtourism can be defined as an abundance of detrimental effects caused by tourism on both local communities and the environment. Overtourism is also described as an excessive increase in visitors, leading to crowding areas, where local residents face the impact of temporary and seasonal peak tourism periods, causing problems in access to services, changes in their lifestyles, and the well-being of residents (Milano et al., 2019). The phenomenon of overtourism is observed in destinations, where the carrying capacity of supply exceeds the limits and is at a critical stage (Sarantakou & Terkenli, 2019). According to the World Tourism Organization (UNWTO, 2018), overtourism is defined as the influence of tourism on a destination, or specific areas within it, that significantly diminishes the perceived quality of life for residents and negatively impacts visitors' experiences (UNWTO, 2018). Peters et al.'s research (2018) on overtourism describes this phenomenon as a scenario where the influence of tourism, in certain areas, surpasses the physical, ecological, social, economic, psychological, and capacity limits. According to the World Travel & Tourism Council (WTTC & McKinsey, 2017), overtourism is turning attractive destinations into victims of their popularity.

Public skepticism is related to the growing recognition that mass tourism has significant costs in cultural, social, and environmental terms. Even more, local economic benefits may be very limited in some niche markets that drive overtourism, such as cruise or bus tourism, while the burden on populations and ecosystems is heavy (Muler et al., 2018). In addition, technological developments are shaping tourism behavior and can therefore lead to undesirable consequences. Social media can make tourists flock to areas that lack the necessary infrastructure, thus increasing the harmful effects of overtourism there (Seraphin et al., 2018). Potentially damaging effects of overtourism include increased costs of living and housing (mainly through the use of sharing economy platforms) and speculation in real estate, congestion of transport infrastructure, alteration of the identity of residents, loss of an authentic character of destination, significant damage to cultural or environmental heritage, or privatization of sites that are supposed to be accessible to the public (Seraphin et al., 2018; Panagiotopoulos & Pisano, 2019). IPK (2017/2018) reports that 25% of international tourists visiting 24 countries believed their chosen destinations suffered from overtourism, with 9% personally affected by it. Seasonality is a significant concern in many Greek tourist destinations, resulting in wage instability and irregular employment patterns for workers. Negative repercussions of overtourism can trigger a chain reaction, altering the entire tourism landscape of a destination (Milano et al., 2019).

Achieving sustainable tourism is an ongoing process, which requires the active involvement of all the necessary stakeholders. Sustainable destination management practices are essential for maintaining high-quality tourist experiences, visitor satisfaction, and the well-being of local communities (UNWTO, 2018). Several strategies can address overtourism effectively. These include revising promotional and marketing strategies to redirect tourists to lesser-known destinations, thus alleviating congestion in popular areas. Imposing fines can serve as an efficient deterrent, with the revenue reinvested in tourism product development, infrastructure improvements, and environmental conservation (Hanna et al., 2018; INSETE, 2021). Additionally, diversifying target segments and distributing visitors spatially and temporally can help manage overtourism. Tailoring experiences and products to different visitor segments, dispersing tourist attractions geographically, and relocating cultural events to less touristic areas facilitate the distribution of tourist flows (Berger, 2018). INSETE (2021) recommends immediate regulatory actions to control traffic and alleviate overcrowding in popular Greek destinations. Measures to regulate accommodation supply through the sharing economy and restrict activities contributing to overtourism are also proposed.

Sharing Economy

The term sharing economy appeared in dictionaries and studies in 2015. In 2015 and onward, it has been shown that this new trend in the economy is the beginning of a new economic activity and is not an ephemeral source of income. It is quite a powerful force both culturally and economically as it revolutionizes both "what" users consume and "how" (Botsman & Rogers, 2010a). An important criterion of the sharing economy is that it allows individuals to generate revenue from assets that are not fully utilized (Zervas et al., 2017). According to Botsman and Rogers (2010b), the sharing economy is an "economic system based on the sharing of underutilized goods or services, free of charge or for a fee, directly by individuals." The sharing economy is described as an "alternative to private ownership that emphasizes both market exchange and gift giving" (Belk, 2007).

In the economic sector, sharing economy companies offer significant job opportunities, enabling extra income through temporary work without full-time commitments. However, this situation has its negatives, as it is easy for this type of work to remain undeclared. This can increase tax evasion and reduce the number of full-time jobs offered by companies. Another important impact of the sharing economy is the changes required in the institutional and legal framework of countries. For instance, many hoteliers complain that they pay exorbitant taxes and are obliged to maintain high-quality standards, while they need to compete with private individuals, who rent out their homes without any quality and tax obligations. The Confederation of National Hotel, Restaurant, and Leisure Associations of the Member States of the European Union (HOTREC, 2016) argues that unfair competition is being created between members of the sharing economy. The study points out that it creates risks to consumer protection and safety. According to Grant Thorton's study (2015), the sharing economy has implications for the functioning of institutions. Due to the absence of an appropriate institutional framework, property owners do not follow a relevant licensing procedure. Nevertheless, they represent a part of the tourism market for which there is no official data on both size and the type of tourism services they offer. Thus, bodies such as the Ministry of Tourism or the Hellenic Chamber of Hotels cannot play their full role, as part of the tourism market is outside the remit. In addition, the sharing economy has an impact on the quality of the services provided, as it raises issues of personal safety and public health. At the same time, public security issues may also arise as the data of visitors are not recorded. Furthermore, the provision of substandard tourism services may have a negative impact on the overall visibility of the country's tourism product, harming the industry's professionals (Grant Thorton, 2015).

HOTREC (2016) proposes actions that aim to make the sharing economy a sustainable and integrated model, ensuring consumer safety. Some of the key measures could be to control the dispersion of private homes for short-term rentals, to impose requirements, to conduct inspections for safety and consumer protection, and to integrate short-term private home rentals into the tourism statistics sector. Furthermore, the Hellenic Ministry of Finance in cooperation with the Ministry of Tourism promotes legislative regulations to address the issue comprehensively. Among the main pillars of the legislative regulation are (*i*) the suppression of underground economy and tax evasion phenomena, ensuring a minimum level of protection for users and third parties, and (*ii*) the clear definition of the boundaries between the business activity developed through the operation of small-scale tourist accommodation and the occasional exploitation by private individuals of additional properties.

Emerging and Secondary Destinations

Tourism destinations are players in the world tourism market. They can be considered "products" in a competitive market. The tourism destinations' market is not only worldwide and competitive but also quite segmented as regards geographical and continental segregations, cultural orientations, and, of course, thematic tourism specializations, such as winter sports, summer activities, clubbing and gambling, folklore and antiquities, natural beauty, or explorations in the jungle and safaris, only to name a few. These segmentations play a specific role in competition and make destinations experience growth, maturity, and decline, while new emerging and/or secondary destinations enter the market and follow similar paths.

Every player in the destination market is a "product" subject to life cycles, as the relevant theory goes. A product's life cycle is the path followed when it is created, developed, enters the market, matures, and finally declines. Levitt (1965) conceptualized the product life cycle in five stages. He pointed out that the product life cycle is hardly used tactically, although it is a widely known concept.

In the initial "Development" stage, the task of bringing a new destination to the market is loaded with risks and often travels in uncertainty. One has to create demand in a new market during this stage. Given the specific attractions of the destination, amenities and facilities have to be designed and materialized, in a customer-oriented development manner. This is a primary condition of sales and profit growth, as it has been demonstrated time after time. Unexpected costs associated with launching insufficiently oriented new products, and frequent fatalities, are common.

During the "Introduction" stage, the destination must be launched in the marketplace, may it be regional or global, which is a critical time in the project's life cycle. During this stage, marketing teams, may it be Governments, Tourism Development Agencies, private tourism chains, or operators, spend a lot of effort and resources on the promotion of the destination's attractions, amenities, and facilities. They should be targeted at building public awareness so that the destination reaches the most suitable audience. While positive results may in some cases be spectacular, introducing a new destination is usually harder than promoting innovations of an existing one (Valeri, 2023, 2024).

The next stages are "Growth" and "Maturity" where a destination is already in the market and tourists have adopted it. Emerging technical problems get solved, demand and visitors increase, and the business starts generating profits at a steady pace. During this process, more direct competitors sooner or later enter the market offering a similar or even improved tourist product. Growth is sustained by adding new qualitative or quantitative features, in the process of refining the tourist product of the destination and by opening new distribution channels. When the destination capacities and attraction can no longer grow, maturity is reached. Production costs may be lower, but demand also becomes more moderate, and pricing has to be more competitive if the destination's appeal is to be sustained.

Emerging destinations may have taken a portion of the market and many tourists are streamed there. The challenge now is to support the destination's competitive position and take the measures necessary to avoid further setbacks. Marketing efforts are focused mainly on highlighting the destination's individuality and preserving quality, instead of creating awareness. Technological innovation is crucial; for example, free and powerful connectivity facilities become an asset and a viability criterion. Environmental respect issues have also become a key in modern tourism. Outdated practices on the above, only to name two, are often a reason for market decline.

Emerging or secondary tourism destinations, namely the ones being in, or near, the second stage of the above-described product cycle can be found worldwide. From Russia to Azerbaijan or Mongolia, from the Faroe Islands to Albania, Montenegro, or alternative Greece, from Morocco to Algeria, from hidden places in Latin America or Africa, from India to Tehran, to Vietnam, to Myanmar, or the Philippines, and hundreds of small hidden destinations worldwide, in all continents, provided that there is peace, respect for the civil rights of residents and visitors, and of course adequate security and health provisions, which we will see next.

Security and Crisis Management

The basic prerequisites for addressing security and crises in a destination could be best met if the government, agencies, and private actors could cooperate on the basis of a coherent and tactical Crisis Management Team (CMT) which would produce a comprehensive Crisis Management Plan (CMP) (Valeri, 2022).

A CMT is a contained team representing local government, local safety and security services and agencies, and local actors and stakeholders, whose task is to manage and address all aspects of a crisis incident by liaising and coordinating the agencies and stakeholders involved. Its composition may include senior management, legal, public relations, and operations specialists, and the range of responsibilities may vary depending on the size of the vicinity of responsibility, may it be a city, a region, or a country, and of course the level of authority it may be endowed with. The CMT's specific tasks and responsibilities are collecting and evaluating information and designing programs of preparations, precautions, and response operations, which lead to the construction of the CMP. The CMT is responsible for managing the crisis and ensuring that the CMP is implemented effectively and reports to the political supervisor, may it be the mayor, the regional governor, or the cabinet minister.

A CMP is an elaborate document outlining the procedures and methods an authority should follow, in response to a critical situation with the potential to harm life or property, reputation, profitability, or just normal running of business. Having a CMP in place, the destination's authorities can minimize the impact of crises by quickly responding and taking appropriate action to address and resolve the situation. The early stages of a CMP include processes of risk assessment and threat intelligence which contribute to early identification of threats. This involves assessing their likelihood and potential impact. The risk assessment process should be ongoing and should receive input from all relevant stakeholders. The CMP should provide for developing an organizational resilience culture to be ready and available long before an incident occurs. It should include procedures for evacuating people, securing facilities, providing rescue and relief, water, food, shelter, transport, and managing the crisis in any suitable way. Those provisions, in turn, drive better response results, leading to better social, financial, and political outcomes. A communication plan would outline how the Team will communicate with services, agencies, and stakeholders before and during a crisis. Regular testing, which can involve simulations of different types of crises, should be conducted to ensure that the Plan is effective.

Crisis management planning helps destinations to improve the well-being and safety of local employees and the general public, but also the safety of visitors and tourists. Another advantage of crisis management planning is that it can help destinations and their visitors and tourists mitigate potential legal exposure in the case of a crisis. Examples of security incidents and crises that could possibly happen in tourism destinations are numerous if not infinite. Only to name a few and some indicative responses would be a task that would exceed this note, but still a concise mention can be attempted.

A crisis management plan should consider actions and/or procedures in cases of: civil unrest, shooter, terrorist, or criminal attacks in premises; violence between tourists amidst alcohol and/or drugs abuse in tourist premises or in open spaces; kidnaping, actions of sexual harassment, robberies, thefts, extortion against tourists and petty crime. Also in cases of fire and/or explosion on premises or within vicinity; release of hazardous materials, causing pollution to the environment and threats to health, wildfires, etc. Action plans should also address natural disasters such as: earthquakes, weather emergencies like heavy snowfall, floods, storms in sea and land, power or water provision interruption. This way, a whole lot of actions can be planned, like, rescue, first aid and relief, sheltering and housing, transfer, hospitalization andcounseling, only to name but a few.

In many of the above incidents, if not in all, tourists and visitors are more vulnerable due to communication problems, lack of local knowledge, networking, lack of private means of transport, and so on; thus, tourism destinations should always prioritize tourists in their CMPs.

Post-COVID and Future Health Issues

COVID-19 has brought the world into a comparatively new and unprecedented turmoil (Abimola et al., 2020). Some 770 million cases have been reported from the beginning of the pandemic in December 2019 to August 2023, while fatalities attributed to COVID-19 have been quite considerable, at 7 million people (WHO, 2023). The Case Fatality Rate in the beginning of the pandemic was up to 10%, in certain countries and places, given testing limitations, but has fallen to about 1%, as of the above numbers, given the mutations of the virus and full availability of testing. Still though, COVID-19 has taken by far a heavier life toll than the previous epidemics of modern times, like the H1N1 influenza, the H5N1 and H7N9 avian (poultry) influenzas, or the Ebola virus (Medical Net, 2023).

However, more important, perhaps, has been the exhibition of unpreparedness by national health systems, as well as government agencies and officials, worldwide, regarding the measures taken including business lockdowns, closing of borders, air traffic blockades, education and sports halts, and so on. Social problems emerged because of the measures, such as depression, increase in household violence, and unemployment. On the other hand, there has been a rapid increase in alternative forms of communication, labor, education, and recreation, mostly based on information technology and advances in connectivity, with mixed economic and social results that remain to be studied in detail.

As countries and organizations adapt to the post-COVID environment, the necessity for developing codes of practice and building up expertise in navigating the universe of security risk management and dealing with health crises has become critical, especially as regards tourism destinations. In the future, perhaps, there should be government regulatory requirements for regions and organizations to have adequate CMPs and CMTs in place, of the kind described previously. However, even without regulatory requirements, responsible tourism destination leaders should strive to manage their risk and resilience by moving forward in an uncertain operational environment, especially in cases of health threats and/or incidents, finding it challenging to schedule the appropriate reactions.

Before the pandemic, governments, public agencies, and companies have been preparing security and facilities management, leaving a significant gap around health risk management. After the pandemic, this issue has entered the risk management agenda, often with indicative statements rather than direct guidelines or compulsory measures. At the European Union level, the European Commission guidelines have an agenda of "Strengthening effectiveness," "Increasing accessibility," and "Improving resilience" (European Commission, 2023). These have even greater importance in tourism-receiving regions, as tourists are more vulnerable as regards health issues.

The various recommendations under the Health Chapter of the European Semester can be summarized as follows: Improve health outcomes by supporting preventive health measures and primary health care. Improve access to health services, by reducing patients' payments. Restore shortages of health professionals. Increase the cost-effectiveness without affecting the quality of the healthcare system. Improve long-term fiscal sustainability of pension and healthcare systems. Address the expected increase in age-related expenditure. Focus investment-related economic policy on health, taking into account regional disparities and ensuring social inclusion (European Union, Health and Food DG, 2023).

On the above framework, national and local governments, regional agencies, and organizations, as well as private stakeholders of the tourism destination regions, should strive to have measurable results leading to building capacity and resilience of the regions as regards health incidents of small scale or massive, as soon as possible, and by all means before the next pandemic.

Conclusions, Challenges, and Policy Implications

Tourism toward 2030 presupposes the transition from spontaneous in targeted development through a methodical, meaningful, and integrated design in a sustainability framework taking into account available resources, goals, and emerging trends. To this purpose, the study provides useful information and further develops Greek tourist products, in terms of destinations and clusters across the country. The phenomenon of overtourism worldwide, although spatially and temporally limited, had led (in the period before the pandemic) some tourist destinations to strain their infrastructure, deterioration of the offered product, and strong reactions against tourism from part of the population. After the experience of the COVID-19 pandemic, people are more aware not only of the need to avoid overcrowded conditions but also of the need for sustainability, a concept that runs counter to the exhaustive use of resources, as is the case in overtourism destinations. Often overtourism is an expression of a failure to manage a destination. In such cases, dealing with the phenomenon requires investment in resource management, possibly to increase the capacity of the destination but also to better manage tourist flows through the heterochronism of tourist flows and the development of products in different locations from the busiest ones. On the other side, with the continuous rise of the sharing economy, regulatory management is considered essential to ensure equal terms of competition and the safety of both citizens and tourists.

Goals set for tourism development at the current juncture have to take into account the satisfaction of tourism, the increased offered options and potentials, the ease that technology offers but also the protection of monuments, infrastructure, and the environment. Competitiveness locally but also in the Mediterranean region along with easy and fast access to information urges the need for the advancement of tourism products. Nevertheless, social and demographic changes as well as climate change differentiate tourists' target groups and experiences. After the growth and maturity stages of a destination, more direct competitors enter the market with a similar or improved product, often adding new features, qualitative or quantitative. The Emerging Destinations are opening new critical distribution channels. Sooner or later, they manage to take a portion of the initial market and tourists are streamed to these emerging destinations. The challenge is to maintain the emerging destinations' quality and safety. For addressing security issues in a tourist destination, the government and stakeholders should form a CMT, to design preparations, precautions, and response operations, namely, the CMP. This document outlines the procedures that authorities should follow in response to a critical situation that has the potential to harm life, property, or business. The impact of crises on business can be minimized by quickly taking appropriate action to address and resolve the situation.

As countries adapt to the post-COVID environment, the necessity for developing health crisis risk management practices becomes critical, especially in tourism destinations. Measures and practices promoting quality, resilience, and cost-effectiveness of the healthcare system, preventive health measures and strengthening of primary health care, increased age-related expenditure, and taking into account regional disparities, should be taken, among others.

Our research results enhance the existing academic literature on the tourism agenda for Greece toward 2030 and provide useful information to policymakers and stakeholders who are involved in its implementation. This research work is original as it examines the future Greek tourism agenda and the opportunity to enhance its competitiveness and structural adaptation. Our findings contribute to the Greek tourism literature by providing valuable insights into the tourism agenda toward 2030, especially since the body of such academic research for Greece has been very limited. However, among the limits of this research is that a quantitative empirical work, like a questionnaire survey, to further enhance the research methodology, is not applied.

References

Abimola, S., Dahake, R., Delamou, A., Ingelbeen, B., Wouters, E., Vanham, G., Van De Pas, R., Dossou, J.P., Ir, P., & Van der Borght, S. (2020). *The COVID-19 pandemic: Diverse contexts; different epidemics – how and why?* Retrieved August 30, 2023, from https://gh.bmj.com/content/bmjgh/5/7/e003098.full.pdf

Belk, R. (2007). Why not share rather than own? *The Annals of the American Academy of Political and Social Science, 611*(1), 126–140.

Berger, R. (2018). *Protecting your city from overtourism: European city tourism study 2018.* Austrian Hotelier Association. file:///D:/roland_berger_european_city_tourism_2018.pdf

Botsman, R., & Rogers, R. (2010a). *What's mine is yours, The rise of collaborative consumption.* HarperCollins.

Botsman, R., & Rogers, R. (2010b). *What's mine is yours: How collaborative consumption is changing the way we live. Business.* HarperCollins. www.wired.co.uk/news/archive/2011-10/13

Development Goals – Journey to 2030, Highlights, UNWTO. (2017). www.e-unwto.org/doi/pdf/10.18111/9789284419401

European Commission. (2023). *Health systems coordination.* Retrieved August 30, 2023, from https://health.ec.europa.eu/eu-health-policy/health-systems-coordination_en

European Union, Health and Food DG. (2023). *European semester: Commission proposes health recommendation.* Retrieved August 30, 2023, from https://ec.europa.eu/newsroom/sante/newsletter-archives/15551

Gansky, L. (2010). *The mesh: Why the future of business is sharing.* Portfolio.

Grant Thorton. (2015). *Operation and impact of the sharing economy in the hotel industry in Greece* [in Greek]. Grant Thorton.

Hanna, P., Font, X., Scarles, C., Weeden, C., & Harrison, C. (2018). Tourist destination marketing: From sustainability myopia to memorable experiences. *Journal of Destination Marketing Management, 9.*

HOTREC. (2016). *2015–2016 annual report.* www.hotrec.eu/wpcontent/customerarea/storage/a2b6d8b8398da4fd62182879a91982f6/0416__Hotrec_Annual_Report_WEB_BAT_-_FINAL.pd

INSETE. (2021). *Greek tourism 2030, megatrends* [in Greek]. INSETE.

IPK. (2018). *Boom or bust? Where is tourism heading – ITB world travel trends report.* ITB. (Original work published 2017).

Koens, K., Postma, A., & Papp, B. (2018). Is overtourism overused? Understanding the impact of tourism in a city context. *Sustainability, 10*(12), 4384. https://doi.org/10.3390/su10124384

Levitt, T. (1965, November). Exploit the product life cycle. *Harvard Business Review.* Retrieved August 30, 2023, from https://hbr.org/1965/11/exploit-the-product-life-cycle

Medical Net. (2023). *How does the COVID-19 pandemic compare to other pandemics?* Retrieved August 30, 2023, from www.news-medical.net/health/How-does-the-COVID-19-Pandemic-Compare-to-Other-Pandemics.aspx

Milano, C., Cheer, J. M., & Novelli, M. (2019). *Overtourism: An evolving phenomenon, overtourism: Excesses, discontents and measures in travel and tourism.* CABI.

Muler Gonzalez, V., Coromina, L., & Galí, N. (2018). Overtourism: Residents' perceptions of tourism impact as an indicator of resident social carrying capacity – case study of a Spanish heritage town. *Tourism Review, 73.*

Panagiotopoulos, A., & Pisano, C. (2019). Overtourism dystopias and socialist utopias: Towards an urban armature for Dubrovnik. *Tourism Planning & Development, 16*(4).

Peters, P., Gössling, S., Klijs, J., Milano, C., Novelli, M., Dijkmans, C. H. S., Eijgelaar, E., & Hartman, S. (2018). *Research for TRAN committee – overtourism: Impact and possible policy responses; policy department for structural and cohesion policies dir.* https://www.europarl.europa.eu/RegData/etudes/STUD/2018/629184/IPOL_STU(2018)629184_EN.pdf

Sarantakou, E., & Terkenli, S. T. (2019). Non-institutionalized forms of tourism accommodation and overtourism impacts on the landscape: The case of Santorini, Greece. *Tourism Planning and Development, 16.*

Schor, J. B. (2015). *Collaborating and connecting: The emergence of the sharing economy.* https://www.academia.edu/13505589/Collaborating_and_Connecting_The_Emergence_of_a_Sharing_Economy

Seraphin, H., Sheeran, P., & Pilato, M. (2018). Over-tourism and the fall of Venice as a destination. *Journal of Destination Marketing & Management, 9.*

UNWTO. (2018). *"Overtourism"? Understanding and managing urban tourism growth beyond perceptions.* UNWTO.

Valeri, M. (2022). *Tourism risk: Crisis and recovery management.* Emerald Publishing.

Valeri, M. (2023). *Tourism innovation in the digital era: Big data, AI and technological transformation.* Emerald Publishing.

Valeri, M. (2024). *Innovation strategies and organizational culture in tourism*: Concepts *and case studies on knowledge sharing.* Routledge Publishing.

WHO. (2023). *World Health Organisation, fact-sheets.* Retrieved August 30, 2023, from www.who.int/news-room/fact-sheets/detail/coronavirus-disease-(covid-19)

WTTC & McKinsey. (2017). *Coping with success – managing overcrowding in tourism destinations.* WTTC & McKinsey.

Zervas, G., Proserpio, D., & Byers, J. W. (2017). The rise of the sharing economy: Estimating the impact of Airbnb on the hotel industry. *Journal of Marketing Research, 54,* 687–705. https://doi.org/10.1509/jmr.15.0204

9 Journeying Beyond Reality

Exploring India's Metaverse Marvel

*Manisha Paliwal, Nishita Chatradhi,
and Marco Valeri*

Introduction

Over the course of the previous two decades, there has been a notable increase in the development of novel and captivating consumer experiences within the context of technological progress. The proliferation of these technologies has had a profound impact on human culture. According to Kaur et al. (2023), the concept of the metaverse has been heralded as a promising new domain that holds significant potential for fostering lucrative business prospects. According to the study conducted by Golf-Papez et al. (2022), the concept of the metaverse is frequently referred to as Web 3.0 in scholarly discourse (Nalbant & Uyanik, 2022; Koo et al., 2022; Yang & Wang, 2023).

In contemporary digital environments known as the metaverse, individuals engage in real-time interactions through the utilization of avatars within interconnected and enduring virtual spaces (Zhu et al., 2022; Belk et al., 2022; Hazy, 2012; Verma & Rao, 2016; Narayanswamy, 2016; Monaco & Sacchi 2023; Kraus et al., 2022). AR/VR technology is exerting a transformative influence on various aspects of everyday life, encompassing communication, entertainment, workforce development, and education. Many of the areas discussed earlier still suffer from a significant dearth of research or empirical foundation (Khalid et al., 2023). In order to effectively and securely create and utilize augmented reality/AR/VR tools as the field continues to expand into new applications and user populations, developers and implementers from various sectors, including private enterprises, government entities, and individuals, will need to possess a strong foundational understanding (Paliwal et al., 2024; Chakravarty et al., 2021).

In several of the categories, there remains a noticeable lack of research and empirical evidence. The metaverse is anticipated to encompass a combination of various features and technologies, such as VR, AR, user-generated games and texts, user interfaces, and digital avatars (Zheng et al., 2022; Kim 2021; Martins et al., 2022). The transitory nature of the metaverse will be influenced to some extent by the utilization and recombination of these characteristics (Bourlakis, 2009; Suanpang et al., 2022). This convergence of technologies within the metaverse is commonly referred to as Web 3.0.

DOI: 10.4324/9781003497004-11

　　The utilization of VR, AR, and other technological advancements has witnessed a notable surge in recent times (Zhu et al., 2022; Belk et al., 2022; Hirsch et al., 2022). The concept of the metaverse, along with its associated technologies such as augmented reality, virtual reality, blockchain, cryptocurrencies, and non-fungible tokens, offers a unique opportunity for a deeply engaging experience that bridges the gap between the physical and virtual realms (Golf-Papez et al., 2022; Hilken et al., 2017). According to recent market research reports, there appears to be a lack of clarity among companies and their customers regarding the concept of the metaverse. According to the study conducted by Golf-Papez et al. (2022), numerous enterprises are adopting a limited viewpoint when it comes to the metaverse, perceiving it solely as a virtual replica of the physical world or akin to the virtual platform Second Life (Valeri, 2021). During the early years of the new millennium, Second Life emerged as a virtual platform that facilitated user interaction through the use of avatars, enabling the creation and exchange of digital goods. According to the study conducted by Kaplan and Haenlein in 2009, it was found that, despite receiving significant attention in certain sectors, Second Life was unable to achieve

Attitudes Toward the Metaverse Among US Adults, Jan 2022

% of respondents

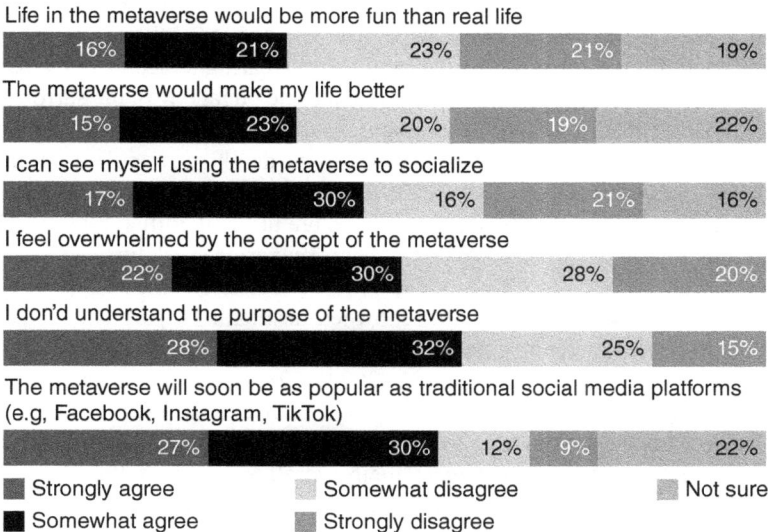

Life in the metaverse would be more fun than real life

| 16% | 21% | 23% | 21% | 19% |

The metaverse would make my life better

| 15% | 23% | 20% | 19% | 22% |

I can see myself using the metaverse to socialize

| 17% | 30% | 16% | 21% | 16% |

I feel overwhelmed by the concept of the metaverse

| 22% | 30% | 28% | 20% |

I don'd understand the purpose of the metaverse

| 28% | 32% | 25% | 15% |

The metaverse will soon be as popular as traditional social media platforms (e.g, Facebook, Instagram, TikTok)

| 27% | 30% | 12% | 9% | 22% |

■ Strongly agree　　　▨ Somewhat disagree　　　▨ Not sure
■ Somewhat agree　　　▨ Strongly disagree

Figure 9.1 Attitude toward metaverse.

Source: eMarketer, Attitude Towards the Metaverse Among US Adults, January 2022

mainstream adoption and remained in the pro- to metaverse stage (Valeri, 2024b, 2023; Jayawardena et al., 2023).

The Metaverse refers to a digital hereein individuals can engage in interactive activities and immerse themselves within a three-dimensional setting (Kaur et al., 2023; Go & Kang, 2023). The composition consists of several strata of digital elements, encompassing VR, AR, and three-dimensional (3D) graphics. The technology exhibits a diverse array of potential applications, one of which includes the field of marketing (Toraman 2022; Um et al., 2022).

Rise of Augmented Reality and Metaverse in Tourism in India

The tourism industry in India holds significant importance for the nation's economy and is experiencing a notable rate of growth. According to the Travel and Tourism Competitiveness Report 2019, India was ranked 34th out of a total of 140 nations in terms of overall competitiveness in the travel and tourism sector. India made significant progress in its positioning, advancing by six places in the 2017 report. The improvement was the most notable among the top 25% of nations in terms of their rankings. As of August 2019, the United Nations Educational, Scientific, and Cultural Organization (UNESCO) has officially recognized approximately 38 World Heritage sites in India. The utilization of computer-generated VR technology in experimental trials is significantly transforming the field of travel and tourism (Shaikh et al., 2018; Kumar & Shekhar, 2020).

In light of this consideration, the Ministry of Tourism, Government of India, has forged a partnership with Outside VR, a travel technology startup, with the aim of enabling individuals to experience virtual travel in India and thereby augment the ongoing Incredible India promotional initiatives. Wearing a racehorse mask while traveling hinders one's ability to fully appreciate and engage with the surroundings and the inherent splendor of the destination. Insufficient awareness regarding travel destinations is a prevalent factor contributing to the decline in popularity of numerous tourist attractions (Chen et al., 2023; Fernandez & Hui, 2022).

The tourism and hospitality sectors in India are utilizing virtual reality technology to augment and enrich the overall visitor experience (Paliwal et al., 2022; Kumar & Shekhar, 2020). Virtual tours provide tourists with the opportunity to engage in immersive experiences of renowned destinations, partake in local cultural activities, and navigate significant landmarks, all in the absence of physical presence (Raj et al., 2023; Samuel et al., 2024). Hotels and resorts are employing this technology as a means to exhibit their amenities and offer virtual tours of their rooms. This technological advancement enables prospective travelers to make better-informed choices and enhances their overall level of satisfaction (Paliwal et al., 2023; Kumar & Shekhar, 2020). The potential integration of the metaverse into the tourism sector presents an opportunity for the country to offer distinctive experiences within the virtual realm of the Indian metaverse. In order to enhance tourist attraction, the implementation of virtual experiences for renowned landmarks such as the Taj Mahal, Red Fort, and Golden Temple holds potential as a viable strategy (Raj et al., 2023).

Prominent urban centers in India, such as Mumbai, Goa, Jaipur, and others, can also be included. Individuals have the ability to acquire knowledge regarding the services offered by a hotel and gain a sense of the atmosphere of the establishment before making a reservation. Prospective travelers have the opportunity to obtain a three-dimensional preview of the offerings and experiences available in this remarkable nation before finalizing their travel arrangements (Sanjeev & Tiwari, 2021). Airfare prices can often be prohibitively high, and even when one manages to secure a more affordable ticket, it is advisable to engage in virtual exploration of the destination in order to maximize the value of one's investment.

According to Global Data's Q1 2021 consumer survey, 43% of global Gen Z participants were influenced by a product or service's digital advancement. This suggests that this demographic is highly likely to like the metaverse in the digital age (Valeri, 2024a; Yadav et al., 2022). It is surprising since most of this group does not remember life before smartphones. Tourism metaverse adoption depends on this cohort.

India is a global technology hub with many software engineers, developers, and startups (Qasim et al., 2023). Due to its rapidly growing technological ecosystem, the nation can help expand the metaverse. Indian tech giants are investing more in AR, VR, and blockchain, the metaverse's building blocks (Paliwal & Chatradhi, 2024; Santus et al., 2023). Tata Consultancy Services, Infosys, and Wipro are pioneers in creating immersive digital experiences using their expertise. About 260 AR/VR startups have emerged in India in recent years. These startups are excelling in AR/VR niches such as education, gaming, and healthcare (Business Today, 2020). Tourism in India's AR/VR landscape is underexplored. Tourism, which could benefit greatly from AR/VR technologies, is underrepresented in AR/VR integration. AR/VR is underutilized in tourism despite its potential (Khan et al., 2022; Gursoy et al., 2022).

Methodology

The study utilizes a case study methodology as its principal approach to achieve its objectives. In order to acquire valuable and in-depth data, the researchers conducted open-ended semi-structured interviews with Mr. Rohit Shrivastaw, the founder and executive director of BharatVerse. The present study can be categorized as descriptive explanatory research, which is distinguished by its utilization of a qualitative methodology. In conjunction with the conducted interviews, the researchers made use of supplementary data derived from diverse sources including government reports, official documents, scholarly articles, published books, and online resources.

The decision to employ a case study methodology was driven by the intention to thoroughly investigate the topic of interest. This methodology facilitated the acquisition of comprehensive and in-depth knowledge pertaining to the subject of investigation. The utilization of an empirical case study approach facilitated a thorough examination of current phenomena, taking into account their real-world contexts, as highlighted by Yin (2014). The selection of this methodology is notably advantageous in augmenting our comprehension of specific phenomena and the complex organizational and political dynamics within society. Consequently, the utilization of a case study as a research instrument proved to be highly advantageous, enabling a comprehensive and intricate approach to the gathering and examination of data, in accordance with Yin's established methodology framework.

BharatVerse – Case Study

The name "BharatVerse" ingeniously merges the concepts of "Bharatvrsh" (an ancient term for India) and "metaverse," indicating its dual nature as a digital nation. Rohit Srivastwa, an accomplished entrepreneur, envisioned BharatVerse as a collaborative endeavor to establish a virtual counterpart of India, allowing individuals to interact with and contribute to a comprehensive representation of the nation. The BharatVerse platform was officially launched in its public beta version on August 15, 2022, showcasing its potential and possibilities. Bharatverse stands as India's first metaverse platform, designed to magnificently showcase the grandeur of Bharat through the application of cutting-edge technologies.

> "BharatVerse is an attempt to bring our glorious Bharat onto the metaverse. In simple words, we envision BharatVerse as a collaboratively built virtual Bharat where every square metre of the real Bharat can be showcased as it exists. We are building the platform on the three-pronged principles of collaboration, creativity, and contribution," Srivastava said.

"BharatVerse" is a metaverse that has been constructed as a virtual representation of India, providing an inclusive platform for individuals who possess a keen interest in exhibiting the diverse aspects of the Indian subcontinent. Indeed, the utilization of wordplay in this context is readily apparent, as it involves a clever play on words involving the terms "BharatVerse" and "metaverse." Individuals who engage in the BharatVerse have the opportunity to obtain entitlements to virtual "Plots" within the BharatVerse platform, which they can utilize to exhibit or generate revenue according to their own discretion. The virtual "Plots" will be linked to real geographical locations, enabling owners to create supplementary products and services that can be integrated into the existing BharatVerse platform.

The main goal of this metaverse is to showcase India's rich history, geography, and culture. The platform uses a variety of products and services in the BharatVerse technology stack to promote local arts and culture. BharatVerse is built on *Collaboration, Creativity, and Contribution*. Collaboration means working with others to achieve a goal. The BharatVerse platform inherently promotes collaboration. The fundamental principle that underlies the BharatVerse project aligns with the core tenets of democracy, namely, "of the people, for the people, and by the people." The concept of creativity refers to the ability to generate novel and valuable ideas and solutions. The primary practical uses of BharatVerse primarily revolve around artistic endeavors, particularly in the realms of virtual tourism and the exhibition of local arts and culture. A proportion of the generated revenue on BharatVerse will be reinvested into the local economy in the physical realm.

The BharatVerse aims to:

- Provide a virtual space to explore every inch of Bharat from the comfort of their homes and offices.
- Provide a virtual platform to showcase the local art and culture.
- Provide a secondary revenue source for BharatVerse participants through an innovative revenue-sharing mechanism.

BharatVerse has a simple mission – to design, develop, and display the best versions of Bharat, i.e., India as it exists or can exist in the future. And, just as it took an entire population to create the Bharat that exists today, we are hoping to enlist the help and support of several people who can come together to make the BharatVerse a success.

Rohit Srivastava, said

BharatVerse envisions creating a digital twin of places of special interest so that the same can be experienced by people across the globe, helping promote places such as RDKM. The metaverse experience will be hosted on Bharat Verse's platform and will be accessible to users through various devices such as head-mounted devices, mobile phones, web browsers etc.

They have developed BharatVerse that annotates the travel destinations in India, highlighting their historical and cultural significance. This initiative aims to raise awareness and promote the preservation of these tourist spots for the benefit of society. With the objective of promoting tourism in India and facilitating convenient travel for global visitors, their proposal entails the development and enhancement of metaverse in tourism. The application serves as a comprehensive guide, offering recommendations for various travel destinations, along with detailed information on their cultural and historical significance. Additionally, the application provides virtual guidance, allowing users to experience the splendor of these travel spots through advanced technologies that offer highly realistic virtual visits.

BharatVerse holds the potential to revolutionize virtual tourism, cultural preservation, and economic growth. By facilitating global access to India's cultural heritage and offering revenue opportunities, it can empower local artisans, entrepreneurs, and enthusiasts. The metaverse's immersive experience, accessible through various devices, promises to engage a diverse audience. BharatVerse exemplifies the harmonious convergence of technology, culture, and community engagement. By establishing a collaborative and creative digital realm, it seeks to replicate and celebrate the multifaceted essence of India.

The interview revealed a growing interest in integrating Metaverse with tourism. The concept allows people to virtually visit tourist destinations from home, providing an immersive and remote travel experience. To improve tourists' VR experiences (Guttentag, 2010), the quality of authenticity is very important. This is because some VR tourism activities may not feel real because the technology used is not very good (Mura et al., 2017). Also, tourists think that physical and sensory participation is an important part of feeling a good level of authenticity in virtual tourism. The development of a tourism-focused Metaverse presents a significant challenge in achieving photorealistic representations. The representation of tourist destinations in content related to tourism should be faithful to the actual physical locations, in contrast to the portrayal of destinations in other virtual reality experiences. One of the primary technical obstacles involves the precise representation of real-life environments and objects, ensuring a high level of visual accuracy.

BharatVerse has accomplished a remarkable feat in the field of virtual tourism by meticulously creating 360-degree photorealistic representations, immersive VR tours, and captivating videos of several iconic heritage destinations throughout India. Among these destinations are Pune, Munnar, Gujarat, Agra, and Varanasi. Some travel destinations are difficult to navigate, which can have a major psychological impact. The application also provides comprehensive information about the specific locations. Virtual museums have revolutionized the way cultural institutions engage with audiences, offering unique opportunities for immersive and interactive experiences.

BharatVerse effectively encapsulates the fundamental aspects of the potential of VR technology, as accurately described by Tussyadiah et al. (2018). Within the context of BharatVerse, individuals are not passive observers, but rather engaged contributors within a dynamic and immersive digital realm. BharatVerse's VR experiences afford users the sensation of being an indispensable component within the virtual realm they navigate. As emphasized by Tussyadiah et al. (2018), BharatVerse utilizes VR devices to promptly identify and monitor user responses and movements. The aforementioned technology exhibits a dynamic capability to adapt to the virtual environment in real time, ensuring a harmonious alignment with the user's actions and emotional states. As a result, individuals experience a highly intricate and dynamic cognitive depiction of their environment, attaining a remarkable degree of engagement and absorption within the simulated realm.

The interview showed that metaverse technology favors animated and computer-generated content. Photorealism in virtual environments is difficult, especially when trying to capture and retain a physical location's imperfections and intricacies. Head-mounted displays like the Meta Oculus help deliver the metaverse experience. However, the high cost of these technologies prevents their widespread use, especially in India. The metaverse's accessibility is a concern. High-end immersive devices provide the best experience, but many people cannot afford them.

The interview examined how shopping mall experience centers could help access metaverse content. Financial sustainability and user willingness to pay for metaverse experiences were key topics. Virtual tourism is in demand, but convincing users to spend money on it is difficult. The Archaeological Survey of India and other government agencies may help integrate historical and cultural sites into the Metaverse. Governments are hesitant to fully adopt this technology. The AR/VR market in India is experiencing substantial growth, particularly in the retail and gaming segments within the consumer space. While these sectors have witnessed significant adoption and innovation in the realm of augmented and virtual reality, it is noteworthy that AR/VR is being extensively utilized in various other industries as well, including education, healthcare, and retail.

Challenges and Opportunities in the Indian Metaverse Industry

a. Challenges

The perception that metaverse experiences are costly and relatively nascent presents a substantial obstacle to their broad acceptance. Investor reluctance to allocate resources toward technology that is perceived as unproven and expensive has impeded the progress of the industry. The private investment predicament in the metaverse industry necessitates a cautious approach for startups and ventures when considering the acceptance of external investors' funds in the absence of a well-defined trajectory toward achieving profitability. Ensuring prudent financial management is imperative in order to mitigate the risks associated with premature scaling or unsustainable growth.

The potential for government support to stimulate the metaverse industry is significant; however, obtaining such support continues to present difficulties. Government initiatives frequently arise as a result of urgent requirements, such as the necessity to promote tourism, which can impose constraints on their extent and scheduling. Despite assertions of broad acceptance, there exist notable obstacles in attaining acceptance for metaverse experiences. Resistance is present among both the general populace and governing bodies, thereby requiring endeavors to persuade stakeholders regarding the advantages and feasibility of the technology.

Limitations of Technology: Presently, metaverse devices, such as AR and VR glasses, are accompanied by certain limitations, encompassing substantial expenses, unwieldiness, and restricted applications beyond the realm of gaming. These factors pose obstacles to extensive adoption, particularly in markets with a strong emphasis on price sensitivity, such as India.

The prioritization of photorealism is essential in metaverse environments in order to effectively deliver immersive experiences. Users have an inherent expectation for virtual environments to faithfully recreate cultural artifacts, historical sites, and other properties in order to augment their level of engagement. In order to promote user investment in AR and VR devices, it is imperative to establish a range of incentives that extend beyond the scope of metaverse tourism. The adoption of devices that provide a variety of applications and serve multiple purposes may be enhanced.

b. Opportunities

Entrepreneurs hailing from various domains, including IT infrastructure and cybersecurity, possess the potential to offer inventive solutions and concepts that can drive the advancement of the Indian metaverse sector (Schindler et al., 2023; Sinha et al., 2024). The possibility of the interviewee's participation as a co-author in an academic study underscores the potential for professionals in the industry to provide valuable real-world insights in the field of metaverse research. Government support presents a significant opportunity, albeit one that is difficult to obtain. Projects initiated by the government that prioritize tourism, cultural preservation, or education have the potential to create favorable conditions for the flourishing of the metaverse industry. The advancement of cost-effective and adaptable AR and VR devices, such as AR glasses, has the potential to foster greater societal acceptance. The increased affordability and expanded functionality of lower-priced devices have the potential to enhance the accessibility of metaverse experiences, enabling a wider range of individuals to partake in such immersive virtual environments.

Implications

The complexities surrounding environmental protection in tourism destinations necessitate a thoughtful and innovative approach to ensure long-term sustainability. While extreme measures such as limiting visitor numbers or curtailing the promotion of tourism products may seem necessary to protect fragile ecosystems, the practical implementation of these ideas presents significant challenges. Many tourism-dependent regions rely heavily on revenue generated from selling tourism products and experiences, making such actions economically unviable. The dilemma lies in the potential exacerbation of economic disparities, particularly in underdeveloped countries or destinations, where tourism serves as a vital lifeline. Restricting tourism development in these regions could lead to increased economic burdens, accentuating the inequality in responsibility for environmental protection and sustainable development. Therefore, addressing the sustainability of tourism destinations in the face of environmental threats demands innovative solutions that strike a delicate balance between conservation and economic well-being. New approaches must be explored to promote responsible tourism practices, encourage eco-friendly initiatives, and implement effective regulations that safeguard the environment without disproportionately burdening economically vulnerable destinations. The pursuit of sustainable tourism is not only an environmental imperative but also a socio-economic one, requiring collaboration among stakeholders to develop comprehensive strategies that ensure the long-term vitality of both tourism destinations and the planet.

Conclusion

India finds itself at the precipice of a transformative metaverse revolution. The nation's potential to establish a presence in the digital frontier is reflected by its dynamic tech industry, entrepreneurial drive, and an increasing number of metaverse startups. Nevertheless, this expedition also presents obstacles pertaining to digital inclusivity, regulation, and societal ramifications that necessitate careful consideration and resolution. As India traverses the metaverse, it possesses the potential to not only influence its own digital trajectory but also make substantial contributions to the worldwide metaverse milieu. The demand for AR/VR in India is robust and extends beyond the realms of edtech, manufacturing, healthcare, and retail. It is poised for expansion into sectors such as travel, hospitality, and media and entertainment. As technology continues to evolve and user expectations grow, the tourism industry could leverage AR/VR to offer immersive and engaging experiences to travelers.

The integration of India with the metaverse is expected to have significant social and cultural ramifications. The concept of the metaverse offers the potential for a virtual environment that enables individuals to establish connections without being constrained by geographical limitations, thereby potentially fostering cultural unity. Nevertheless, this development also gives rise to inquiries regarding the safeguarding of data privacy, the promotion of digital inclusivity, and the increasing convergence of the tangible and digital realms. Ensuring digital inclusion poses a significant challenge for India in its pursuit of the metaverse. The digital divide continues to pose a substantial obstacle, as a considerable number of individuals in India lack access to high-speed Internet and state-of-the-art devices. The act of bridging this divide will be of utmost importance in the pursuit of democratizing the metaverse and guaranteeing inclusive access to its potential for all strata of society.

Acknowledgments

We express our gratitude to Mr. Rohit Srivastava, Founder of BharatVerse, for his tremendous support in the completion of this case study.

References

Belk, R., Humayun, M., & Brouard, M. (2022). Money, possessions, and ownership in the Metaverse: NFTs, cryptocurrencies, Web3 and Wild Markets. *Journal of Business Research, 153*, 198–205.

Business Today. (2020). Hotel sector to witness 50% loan defaults post RBI's August 31 deadline. *Business Today*. Retrieved September 8, 2023, from www.businesstoday.in/current/economy-politics/hotel-sector-to-witness-50-loan-defaults-post-rbi-august-31-deadline/story/408063.html

Bourlakis, M., Papagiannidis, S., & Li, F. (2009). Retail spatial evolution: paving the way from traditional to metaverse retailing. *Electronic Commerce Research, 9*, 135–148. https://doi.org/10.1007/s10660-009-9030-8

Chakravarty, U., Chand, G., & Singh, U. N. (2021). Millennial travel vlogs: Emergence of a new form of virtual tourism in the post-pandemic era? *Worldwide Hospitality and Tourism Themes, 13*(5), 666–676. https://doi.org/10.1108/WHATT-05-2021-0077

Chen, S., Chan, I. C. C., Xu, S., Law, R., & Zhang, M. (2023). Metaverse in tourism: Drivers and hindrances from stakeholders' perspective. *Journal of Travel & Tourism Marketing, 40*(2), 169–184.

Fernandez, C. B., & Hui, P. (2022). *Life, the metaverse and everything: An overview of privacy, ethics, and governance in metaverse*. http://arxiv.org/abs/2204.01480

Go, H., & Kang, M. (2023). Metaverse tourism for sustainable tourism development: Tourism agenda 2030. *Tourism Review, 78*(2), 381–394.

Golf-Papez, M., Heller, J., Hilken, T., Chylinski, M., de Ruyter, K., Keeling, D. I., & Mahr, D. (2022). Embracing falsity through the metaverse: The case of synthetic customer experiences. *Business Horizons*, *65*(6), 739–749.

Gursoy, D., Malodia, S., & Dhir, A. (2022). The metaverse in the hospitality and tourism industry: An overview of current trends and future research directions. *Journal of Hospitality Marketing & Management*, *31*(5), 527–534.

Hazy, J. K. (2012). Leading large: Emergent learning and adaptation in complex social networks. *International Journal of Complexity in Leadership and Management*, *2*(1–2), 52–73. https://doi.org/10.1504/IJCLM.2012.050395

Hilken, T., de Ruyter, K., Chylinski, M., Mahr, D., & Keeling, D. I. (2017). Augmenting the eye of the beholder: exploring the strategic potential of augmented reality to enhance online service experiences. *Journal of the Academy of Marketing Science*, *45*, 884–905. https://doi.org/10.1007/s11747-017-0541-x

Hirsch, L., George, C., & Butz, A. (2022). Traces in virtual environments: A framework and exploration to conceptualize the design of social virtual environments. *IEEE Transactions on Visualization and Computer Graphics*, *28*(11), 3874–3884. https://doi.org/10.1109/TVCG.2022.3203092

Jayawardena, C., Ahmad, A., Valeri, M., & Jaharadak, A. A. (2023). Technology acceptance antecedents in digital transformation in hospitality industry. *International Journal of Hospitality Management*, *108*, 103350.Kaplan, A. M., & Haenlein, M. (2009). The fairyland of second life: Virtual social worlds and how to use them. *Business Horizons*, *52*(6), 563–572. https://doi.org/10.1016/j.bushor.2009.07.002

Kaur, J., Mogaji, E., Paliwal, M., Jha, S., Agarwal, S., & Mogaji, S. A. (2023). Consumer behavior in the metaverse. *Journal of Consumer Behaviour*, 1–19. https://doi.org/10.1002/cb.2298Khalid, R., Hamid, A. B., Raza, M., Promsivapallop, P., & Valeri, M. (2023). Innovation and organizational learning practices in tourism and hospitality sector: A gender-based perspective. *European Business Review*. https://doi.org/10.1108/Ebr-09-2022-0191

Khan, S., & Freeda Maria, S. M. (2022). What innovations would enable the tourism and hospitality industry in India to re-build? *Worldwide Hospitality and Tourism Themes*, *14*(6), 579–585.

Kim, J. (2021). Advertising in the metaverse: Research agenda. *Journal of Interactive Advertising*, *21*(3), 141–144.

Koo, C., Kwon, J., Chung, N., & Kim, J. (2022). Metaverse tourism: Conceptual framework and research propositions. *Current Issues in Tourism*, 1–7.

Kraus, S., Kanbach, D. K., Krysta, P. M., Steinhoff, M. M., & Tomini, N. (2022). Facebook and the creation of the metaverse: Radical business model innovation or incremental transformation? *International Journal of Entrepreneurial Behaviour and Research*, *28*(9), 52–77. https://doi.org/10.1108/IJEBR-12-2021-0984

Kumar, S., & Shekhar. (2020). Digitalization: A strategic approach for development of tourism industry in India. *Paradigm*, *24*(1), 93–108.

Martins, D., Oliveira, L., & Amaro, A. C. (2022). From co-design to the construction of a metaverse for the promotion of cultural heritage and tourism: The case of Amiais. *Procedia Computer Science*, *204*, 261–266.

Monaco, S., & Sacchi, G. (2023). Travelling the metaverse: Potential benefits and main challenges for tourism sectors and research applications. *Sustainability*, *15*(4), 3348.

Mura, P., Tavakoli, R., & Pahlevan Sharif, S. (2017). 'Authentic but not too much': exploring perceptions of authenticity of virtual tourism. *Information Technology & Tourism*, *17*(2), 145–159. https://doi.org/10.1007/s40558-016-0059-y

Nalbant, K. G., & Uyanik, Ş. (2022). A look at the new humanity: metaverse and metahuman. *International Journal of Computers*, 7–13.

Narayanswamy, R. (2016). Leadership is not a destination but a place to come from Gandhi's contribution to evolutionary excellence. *International Journal of Complexity in Leadership and Management*, *3*(4), 278–283. https://doi.org/10.1504/IJCLM.2016.087151

Paliwal, M., & Chatradhi, N. (2024). AI in market research: Transformative customer insights – a systematic review. In M. Rafiq, M. Farrukh, R. Mushtaq, & O. Dastane (Eds.), *Exploring the intersection of AI and human resources management* (pp. 231–255). IGI Global. https://doi.org/10.4018/979-8-3693-0039-8.ch012

Paliwal, M., Chatradhi, N., Singh, A., & Dikkatwar, R. (2022). Smart tourism: Antecedents to Indian traveller's decision. *European Journal of Innovation Management, ahead-of-print*(ahead-of-print). https://doi.org/10.1108/EJIM-06-2022-0293

Paliwal, M., Chatradhi, N., Tripathy, S., & Jha, S. (2023). Growth of digital entrepreneurship in academic literature: A bibliometric analysis. *International Journal of Sustainable Development and Planning, 18*(6), 1929–1942. https://doi.org/10.18280/ijsdp.180629

Paliwal, M., Dikkatwar, R., Chatradhi, N., & Valeri, M. (2024). Evolution of research in knowledge management and competitive advantage. In M. Valeri (Ed.), *Knowledge management and knowledge sharing: Contributions to management science*. Springer. https://doi.org/10.1007/978-3-031-37868-3_1

Qasim, D., Shuhaiber, A., Bany, M. A., & Valeri, M. (2023). E-entrepreneurial attitudes and behaviours in the United Arab Emirates: An empirical investigation in the digital transformation era. *European Journal of Innovation Management*. https://doi.org/10.1108/Ejim-09-2022-0461

Raj, S., Sampat, B., Behl, A., & Jain, K. (2023). Understanding senior citizens' intentions to use virtual reality for religious tourism in India: A behavioural reasoning theory perspective. *Tourism Recreation Research*, 1–17.Samuel, A., Paliwal, M., Saini, D., & Pooja. (2024). Metaverse: Perception and awareness among millennial generation. *2024 IEEE International Conference on Interdisciplinary Approaches in Technology and Management for Social Innovation (IATMSI), Gwalior, India*, 1–6. https://doi.org/10.1109/IATMSI60426.2024.10503183

Sanjeev, G. M., & Tiwari, S. (2021). Responding to the coronavirus pandemic: Emerging issues and challenges for Indian hospitality and tourism businesses. *Worldwide Hospitality and Tourism Themes, 13*(5), 563–568.

Santus, K., Nafi, S., Mallik, N., & Valeri, M. (2023). Mediating effect of emotional intelligence on the relationship between employee job satisfaction and firm performance of small business. *European Business Review, 35*(5), 624–651. https://doi.org/10.1108/Ebr-12-2022-0249

Schindler, J., Kallmuenzer, A., & Valeri, M. (2023). Entrepreneurial culture and disruptive innovation in established firms: How to handle ambidexterity. *Business Process Management Journal, 30*(2), 366–387. https://doi.org/10.1108/Bpmj-02-2023-0117

Shaikh, S., Bokde, K., Ingale, A., & Tekwani, B. (2018). Virtual tourism. *International Research Journal of Engineering and Technology, 5*(4), 2044–2046.

Sinha, M., Shekhar, & Valeri, M. (2024). How does entrepreneurship education promote innovation and creativity? Insights from literature review. *International Journal of Technology Enhanced Learning, 16*(1), Forthcoming Article. https://doi.org/10.1504/Ijtel.2023.10055678

Suanpang, P., Niamsorn, C., Pothipassa, P., Chunhapataragul, T., Netwong, T., & Jermsittiparsert, K. (2022). Extensible metaverse implication for a smart tourism city. *Sustainability, 14*(21), 14027.

Toraman, Y. (2022). User acceptance of metaverse: Insights from technology acceptance model (TAM) and planned behavior theory (PBT). *EMAJ: Emerging Markets Journal, 12*(1), 67–75.

Tussyadiah, I. P., Wang, D., Jung, T. H., & Tom Dieck, M. C. (2018). Virtual reality, presence, and attitude change: Empirical evidence from tourism. *Tourism Management, 66*, 140–154.

Um, T., Kim, H., Kim, H., Lee, J., Koo, C., & Chung, N. (2022, January). Travel Incheon as a metaverse: Smart tourism cities development case in Korea. *In ENTER22 e-tourism conference* (pp. 226–231). Springer International Publishing.

Valeri, M. (2021). *Organizational studies: Implications for the strategic management.* Springer. ISBN: 978-3-030-87147-5. https://doi.org/10.1007/978-3-030-87148-2

Valeri, M. (2023). *Tourism innovation in the digital era: Big data, AI and technological transformation.* Emerald Publishing.

Valeri, M. (2024a). *Knowledge management and knowledge sharing: Business strategies and an emerging theoretical field.* Springer.

Valeri, M. (2024b). *Innovation strategies and organizational culture in tourism: Concepts and case studies on knowledge sharing.* Routledge Publishing.

Verma, P., & Rao, M. K. (2016). Authentic leadership approach for enhancing innovation capability: A theoretical investigation. *International Journal of Complexity in Leadership and Management, 3*(4), 284–300. https://doi.org/10.1504/IJCLM.2016.087114

Wei, W. (2024). A buzzword, a phase or the next chapter for the Internet? The status and possibilities of the metaverse for tourism. *Journal of Hospitality and Tourism Insights, 7*(1), 602–625. https://doi.org/10.1108/JHTI-11-2022-0568

Yadav, H., Manisha, P., & Nishita, C. (2022). Entrepreneurship development of rural women through digital inclusion: examining the contributions of Public Programs. In *Inclusive Businesses in Developing Economies: Converging People, Profit, and Corporate Citizenship,* (pp. 287–309). Cham: Springer International Publishing.

Yang, F. X., & Wang, Y. (2023). Rethinking metaverse tourism: A taxonomy and an agenda for future research. *Journal of Hospitality & Tourism Research.* https://doi.org/10.1177/10963480231163509

Yin, R. K. (2014). *Case study research: Design and methods.* Sage.

Zheng, K., Kumar, J., Kunasekaran, P., & Valeri, M. (2022). Role of smart technology use behaviour in enhancing tourist revisit intention: The theory planned behaviour perspective. *European Journal of Innovation Management.* https://doi.org/10.1108/Ejim-03-022-0122

Zhu, R., & Yi, C. (2024). Avatar design in Metaverse: The effect of avatar-user similarity in procedural and creative tasks. *Internet Research, 34*(1), 39–57.

Zhu, T., Zhang, L., Zeng, C., & Liu, X. (2022). Rethinking value co-creation and loyalty in virtual travel communities: How and when they develop. *Journal of Retailing and Consumer Services, 69*, 103097.

10 Digital Marketing Metaverse for Restoration of Tourism Performance After COVID-19 Business Crisis in Sub-Saharan African Countries

Experience From Tanzania

Alberto Gabriel Ndekwa

Introduction

Many countries in the world consider and regard the tourism sector as one of the main contributors to national economic development. According to Lincoln (2013), tourism is regarded to be a powerful economic and social good, creating employment and wealth and a promoter of international collaboration and friendship. Most policy analysts concur that tourism is one of the world's fastest-growing industries that can contribute to and promote a nation's sustainable development goals (Ndekwa & Katunzi, 2016). Many scholars (Hettiarachchi, 2019; Wonbera, 2019) have cited that tourism is a major source of socioeconomic transformation of a nation which includes financial gains, cross-cultural exchanges, employment opportunities, conservation of historic sites, and improvement of infrastructures.

In Africa, the tourism sector has been identified as one of the main contributors of national changes in lifestyles, community mindset change, and poverty alleviation (Ramukumba, 2019; Wonbera, 2019). Ndekwa (2017) found that tourism helps to alleviate poverty as the income earned in most cases helps family units to meet their basic needs of life. In sub-Saharan Countries, Gnanapala and Sandaruwani (2016) observed that accruals from the tourism industry do contribute significantly to gross domestic product, GDP, and the general improvement of people's lives in sub-Saharan African countries. Odhiambo (2021) found that when the number of tourist arrivals is used as a proxy, the results show that an increase in tourism development consistently leads to increases in household welfare, hence, a decrease in abject poverty. Hence, the promotion of tourism should in all cases focus on innovative improvement of law and order and security of tourist destinations as well as support services.

Even so, the outbreak of COVID-19 during the latter half of the second decade of the 21st century seriously affected the development and role of the tourism industry in many countries in Africa and elsewhere (Ndekwa, 2021a, 2021b). Similarly, cumbersome and bureaucratic immigration procedures and the introduction of many social restrictions resulting from COVID-19 in many nations seriously

DOI: 10.4324/9781003497004-12

affected the smooth flow of tourists around the world. For example, since February 2020, that is after the onset of the COVID-19 pandemic, the number of foreign tourists in each tourism destination in the world decreased drastically, and the peak occurred in April 2020.

The COVID-19 pandemic resulted in a significant drop in tourism worldwide. In some countries, the intensity of the drop in the tourism business was attributed to the baseline economic status and hence poverty level, social, cultural, and in some cases faith-based characteristics of the tourist destination. Most small tourism enterprises failed to survive amidst the bad effects of the antagonistic forces of the business and the effects on the small enterprises contributed significantly to the poor and discouraging effects on the whole of the tourism business sector. Asongu and Odhiambo (2019), cited in Odhiambo (2021), argued that sustainable tourism business is critical in ensuring a constant inflow of income for social and economic development, especially for economies, such as the island states of Zanzibar, Mauritius, and Comoro that depend substantially on the tourism industry for their overall social and economic development and prosperity.

In the effort to restore the earlier performance level of tourism after it was harmed by COVID-19, various strategies have been initiated by countries in the endeavor to regain the traditional performance of tourism markets and operations in general. The strategies include the adoption and use of COVID-19 vaccines as recommended by the World Health Organization and indeed other concoctions and treatment or preventive methods all intended to protect nationals as well as visitors including tourists. Consequently, within a decade, consistency studies (Gvaramadze, 2022; Kavenuke et al., 2017; Ndekwa, 2015; García & Navarrete, 2022) have shown that several technologies and treatment and preventive medicine methods were proposed and used to enhance the health of the people and to rescue the tourism business and related markets. However, it is evidenced that the results of the adoption of these technologies and other preventative and treatment measures have not been able to restore the performance of the tourism sector.

Recently digital marketing metaverse as a strategy has been acknowledged as one of the primary technologies for the restoration of performance of the tourism sector after it was hit by COVID-19. Scholars (Ndekwa, 2023; Buhalis et al., 2023) assert that the key to the restoration of tourism business performance and creative economic actors to survive during and after a pandemic is to adopt and adapt, innovate, and collaborate through digital marketing metaverse. Buhalis et al. (2023) pointed out that digital marketing in the metaverse age is seen to be the main strategy to encourage the restoration and achievement of tourism performance.

The importance of digital marketing is characterized by the application and use of smartphones and social media incorporated into organizational management models. The move has resulted in the application of metaverse to enable changes in the way businesses operate globally. Ndekwa (2021a) recommends that the business sector should design and implement more mobile marketing platforms to offset emerging business crises in the event of a health pandemic. Tourism business and COVID-19 are not easy to separate and dissociate from emerging technologies, especially digital marketing metaverse. Digitization of marketing has caused the business ecosystem to change the way businesses compete (Kannan & Li, 2017; Ndekwa & Katunzi, 2016). Thus, digital marketing metaverse has become one of the strategic areas to engage in for the restoration of tourism businesses and tourist destinations all over the world. Relevant marketing models advocate the use of digital marketing techniques as suggested by different scholars who acknowledge the importance of digital marketing metaverse for the restoration of tourism performance with a focus on improving the characteristics, especially the competitiveness of tourist destinations.

Despite the acknowledgment that metaverse promotes tourism performance, it is well documented that digital metaverse initiatives in the tourism business have not achieved the expected performance goal in the tourism sector. For example, Desai (2019) found that the performance of the tourism business has not improved even after the adoption of digital tools. On the other hand, Velentza and Metaxas (2023) established that many initiatives of tourism businesses have not yet made sufficient use of digital media as a strategic marketing move to cushion the ill effects caused by COVID-19. Studies conducted in Tanzania by (Ndekwa, 2014; Ndekwa & Katunzi, 2016) before the onset of COVID-19 have indicated how the application of ICT in small tourism enterprises was still low despite the promise that the initiative would help to revive small-, medium-, and big-size businesses.

More recent studies (Ndekwa, 2021a, 2023) have indicated a gap of concern between the application of digital marketing and the performance of the tourism business. Other scholars have indicated that hospitality practices worldwide have a standardization gap, especially between local and international tourism enterprises, and more so when it comes to digital business practices (Bhandari & Sin, 2023). It was observed that even before the onset of COVID-19, some nations had already started mainstreaming modern technologies and digital assistants to promote tourism.

Multiple distribution channels for available tourism destinations and attractions were already in place in some countries even before the outbreak of COVID-19. Despite such initiatives, it did not emerge clearly whether the use of digital marketing metaverse could potentially help to restore and promote the performance of tourism businesses or not. This study was initiated in an effort to fill such a gap by analyzing the role and contributions of digital marketing metaverse in restoring the earlier high performance of the tourism business and redressing the business crises in sub-Saharan countries after the end of the COVID-19 pandemic.

Literature Review

Origin of the Concept of Digital Marketing Metaverse

Metaverse is an immersive universe that mixes virtual reality and augmented reality in which users are embodied by avatars and navigate virtual spaces. According to Zakarneh et al. (2014), about 30 years ago, Neal Stephenson developed the term "Metaverse" in his novel Snow Crash. He coined the phrase "metaverse" to describe a technology he conceptualized that combines *virtual reality, augmented reality*, and a *social network*. This development was supported by Stephenson who pointed out that metaverse is a new form of the Internet where augmented reality, virtual reality, blockchain, 5G, and gaming platforms converge as a metaverse. Metaverse, therefore, is the networking and computing of three-dimensional virtual worlds based on anthropomorphized social interaction that combines physical reality and digital through extended reality (XR). It is a technology that incorporates digital and physical elements in varying degrees, namely, AR, mixed reality (MR), and VR.

To better understand the role and contribution of metaverse in marketing, one must first grasp metaverse as a concept and its stakeholders. Metaverse is an XR where physical and digital worlds merge to influence the way people shop, socialize, learn, play, work, and communicate with each other. Metaverse, with its various features and tools, is already empowering brands to deliver offerings that are impossible in the real world. As the laws of nature do not apply to the virtual world, marketers can be imaginative and creative in providing unique products beyond the real world. The virtual nature of metaverse enables brands to interact with a wide range of consumers but with a higher level of immersion.

The graphical representation above provides that in the context of mixed reality (MR), social media connects with the unique capabilities of VR and AR through immersive technology. Virtual media is essential in articulating creative future visions and innovative solutions to complex problems through immersive technologies and this is essential for future metaverse applications

Digital Marketing

According to Chaniago and Arief (2022), digital marketing comprises different channels used by marketers to promote their products and services. According to Ponde and Jain (2019), digital marketing includes marketing through social media, video, mobile devices, search engines, email, affiliate networks, online advertising, word of mouth, search engine optimization, website, google analytics, and text and multimedia content creation. Other scholars such as Buhalis et al. (2023) have defined digital marketing in terms of marketing of any product or service on a digital platform. The authors added that digital marketing is a form of direct marketing that links consumers with sellers electronically using interactive technologies such as emails, websites, online forums and newsgroups, interactive television, mobile communications, and so on. On a related perspective, Rahardja (2022) asserts that digital marketing involves how to market a brand or products using digital media including television, radio, Internet, mobile phone, social media, and various other related digital media where techniques of Internet marketing are included. In this study, digital marketing is considered any Internet technology that has the ability to converge physical setting and virtual setting in a tourism business environment.

Review of Resource Base View Theory

This theory was originally proposed in 1984 by Birger Wernerfelt and later was further developed by Jay B. Barney and other scholars. According to Barney (1991), the resource-based view theory argued for the development of competitive advantage in business through a function of available resources. The theory contends that strategic resources provide an organization with a golden opportunity for competitive advantages over its rivals. In this study, it is observed that metaverse is a resource, and when applied in the tourism sector, especially to small enterprises, it can restore business performance.

This means that competitive advantage in business can be observed when distinctive technological resources such as metaverse are applied in any business firm or industry and constantly reviewed and updated to maintain and increase sector performance. In this study, digital marketing metaverse is regarded as a distinctive resource in tourism with the potential to restore business performance after the scourge of COVID-19.

The suitability of resource-based view in promoting business performance is observed in prior studies (Ndekwa, 2023, 2021a). For example, Ndekwa (2021a) applied this theory to link mobile marketing technological resources and business performance during the COVID-19 business crisis with a focus on tourism. He found a significant impact of mobile marketing on the performance of business in tourism. Given this justification, the application of the resource-based view theory will help in supporting the analysis of how the application of metaverse resources could restore business performance in tourism in sub-Saharan African countries and more specifically in Tanzania.

Empirical Literature Review

Israilov (2021) conducted a study on the link between digital marketing tools and the performance of business in tourism hospitality services. The findings demonstrated that digital marketing has a significant influence on decision-making relating to the potential inflow of tourists which has a direct impact on the occupancy and performance of hotels in the tourism sector. This study focused on how digital marketing can induce decisions that are supportive of and encouraging tourists which in turn enhance business performance in the sector. The study was limited to tourist-related decisions rather than priority over other aspects of digital marketing for the tourism sector.

This calls for further studies to investigate the global perspective of digital marketing metaverse on the performance of the tourism business sector. Chaniago and Arief (2022) found and concluded that the use of digital marketing in the tourism industry to enhance the digital economy plays a critical importance in increasing business in tourism. They added that the use of digital marketing in the area of tourism will not only change the industrial paradigm but also improve how to communicate, shop, transact, and the general lifestyles of people.

In a related perspective, Adel (2023) in his study on the influence of metaverse on interactive experiences of tour operators found that the metaverse has the potential to transform the conduct and trends of business in the tourist industry by enhancing the trends in relating experiences and creating lasting memories for visitors of tourist destinations. However, this study focused only on tourist experiences which lacked how metaverse could help to transform performance trends in the tourism industry and business.

In their study, Wibawa et al. (2022) focused on digitalization of tourism market by evidencing how the application of digital marketing supports sustainable tourism performance in Indonesia. They elaborated that the digital marketing model can make it easier for tourists to find tourist villages in Bangli Regency. They added that through digital marketing, it can make it easier for tourists to carry out tourism activities with an automatic system and multi-language presence. However, this study was carried out in a small area of Indonesia known as Bangli Regency. It is not clear how the findings from this study could provide good evidence for the whole country of Indonesia and outside Indonesia. This is because it is argued that tourism has elements of heterogeneity and is in many cases concentrated in some areas with unique tourism attractions.

Further studies are called for a larger area like Sub-Saharan Africa to provide more valid and rational evidence. Razouk (2023) found and concluded that digital marketing can optimize outreach efforts, increase customer engagement, and contribute to the industry's recovery and growth after the effects of the COVID-19 pandemic. Nassima and Insaf (2023), who carried out a study on e-tourism in Algeria, found that by adopting ICT solutions, particularly in the area of marketing, tourism stakeholders can respond effectively to market demands, improve consumer experiences, and gain competitive advantage in the industry and business.

Based on the theoretical and empirical review earlier, in this book chapter, the author posits the hypothesis that: *Digital marketing metaverse has a significant restorative influence on tourism business performance.*

Conceptual Framework

Based on the theory of resource-based view and empirical evidence, as unearthed in the literature, the researcher drew the following conceptual framework in which digital marketing metaverse is proposed as a means to restore tourism business performance. In this book chapter, the conceptual framework is presented in Figure 10.1.

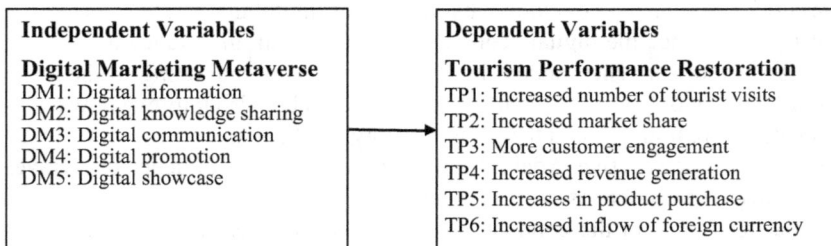

Independent Variables	Dependent Variables
Digital Marketing Metaverse DM1: Digital information DM2: Digital knowledge sharing DM3: Digital communication DM4: Digital promotion DM5: Digital showcase	**Tourism Performance Restoration** TP1: Increased number of tourist visits TP2: Increased market share TP3: More customer engagement TP4: Increased revenue generation TP5: Increases in product purchase TP6: Increased inflow of foreign currency

Figure 10.1 Proposed research model.

Methodology

Research Design and Approach

Both deductive and inductive research approach guided this study. According to Saunders et al. (2012), the deductive research approach has the ability to collect statistical data for the assessment of causal relationships, if any, of the proposed hypothesis. In this study, the researcher aimed to collect statistical data for use in his hypothesis testing on the role of digital marketing metaverse in the restoration of performance trends of tourism business performance after the crisis caused by COVID-19. Inductive research approach was also used.

Inductive research approach on its part produces discoveries that cannot be obtained by using statistical procedures or other means of quantification (Saunders et al., 2012; Creswell, 2014). It is a method that serves to describe an idea of the object under study through data or samples that have already been collected from direct observation. Inductive research approach was applied in this study to collect data that were rich in contextual in-depth information to elaborate the findings obtained from the deductive research approach altogether on the contribution of digital marketing to the restoration of performance of tourism business.

In a related perspective, a cross-sectional research design was applied to enable the researcher to collect data at one time to better understand the metaverse phenomenal on tourism business performance. The epistemological assumptions of the cross-sectional research design in this study are termed as the ability to capture data at one point of time to better acquire knowledge and reality about the role of digital marketing in the restoration of performance levels for tourism business during the post-COVID-19 crisis era.

Study Area

Data for this study were collected in sub-Saharan African countries with a particular focus on Tanzania. Tanzania was selected based on the country's endowment with a full range of tourism attractions. Many small-tourist enterprises are motivated to improve their performance which in turn improves the overall performance of the business sector. Tanzania has had initiatives for the installation of ICT infrastructure which have led to and resulted in several digitalization services also available in the tourism sector despite the fact that small tourist enterprises are passive users of digital tourism tools. Furthermore, Tanzania has many tourism economic activities that small tourist enterprises engage in and which help to build evidence to support the suitability of digital technologies in the restoration of performance of tourism business.

Therefore, Tanzania is and remains a potential contributor to trends in the tourism business performance in sub-Saharan Africa and qualifies for the collection of data needed for this study altogether to discern the understanding and contribution of digital marketing in the restoration of performance level of tourism marketing business performance. Hence, the collection of data from this part of Africa was considered helpful in establishing evidence to support how modern digital metaverse technologies can help promote business performance.

Population and Sampling Design

Study Population

According to Creswell (2014), a population is any collection of individuals that are the focus of a certain survey and are related in some ways. In this study, the population comprised the owners of small tourist enterprises operating in sub-Saharan Africa, especially Tanzania's tourism sector. The group of people involved was chosen focusing on owners of small tourist enterprises. Such people have significant influence in decisions regarding the application of digital marketing metaverse in their enterprises. Thus, the data for this study were collected from a qualified population of owners of small enterprises in tourism and engaged in assessing the influence of digital marketing metaverse on the tourism business sector.

Sampling Technique and Procedure

Systematic sampling was applied in picking a sample of respondents who were owners of tourist enterprises. This technique was considered suitable due to its ability to reduce bias based on the homogeneity of the sample contrary to selecting a sample from a heterogeneous population. In each case, a respondent who is the owner of the small tourist enterprise was selected using an interval case of second during data collection. The owners of small tourist enterprises in tourism were seen to be heterogeneous and did not have the same information base on the role of digital marketing metaverse in restoring the performance of tourism business in the country and hence in sub-Saharan Africa. Systematic sampling was therefore considered relevant for the respondents during data collection.

Sample Size

The relevance and responsiveness of quantitative analysis of results in a study are affected by the sample size of the respondents. However, scholars such as Cochran provide some assumptions needed in calculating accurate numbers of respondents for the reliability of the results. According to Cochran, cited in Yamane, the sample size for a 95% confidence level and a p-value of 0.05 is calculated using the following formula:

$$n = \frac{N}{1 + N(e^2)}$$

N is the population size of a finite population and e is the degree of precision. In this study, the definite number of the population (N) was 632 with a precision(e) of 5%. Using p = 0.5 and a 95% confidence interval, the sample size was

$$n = 632/(1 + 632(0.05^2))$$

n = 632/2.58

n = 244

The researcher used partial least square structural equation modeling (PLS-SEM) to analyze the data. Performance estimators of PLS-SEM are not affected by a small or large sample in producing long-lasting results, but rather PLS SEM tends to enhance sampling distribution to approach normality. Hence, a sample size of 244 as in this study was considered sufficient to produce stable and valid results.

Data Collection Tools

A survey questionnaire and interview schedule were used for primary data collection in this study. The applicability of a survey-structured questionnaire was considered useful due to its ability to capture statistical data for analysis in discerning the significance of influence of digital marketing metaverse in the restoration of performance levels of tourism business after the COVID-19 pandemic. The structured questionnaire was designed with two sections. Section A solicited demographic data with three questions relating to gender, age, and business experience, and section B comprised 11 questions in two themes, one being restoration of performance of tourism business and the second being on digital marketing metaverse.

The participants in the study were requested to respond to the questions using a Likert scale with a maximum score of 5 – "1" standing for strongly disagree, "5" for strongly agreeing, and "3" for uncertain. According to Saunders et al. (2012), questionnaires are valuable for quantitative data collection in studies since they enable researchers to gather organized data for statistical analysis and hypothesis testing. Even so, questionnaires do not have the ability to capture contextual and in-depth information to help in profiling a phenomenal and enhance better understanding. In this case, qualitative data were used to address the need to capture contextual data and in-depth information.

The technique for qualitative data collection was interviews. Interviews were used to collect in-depth information with real life setting to bolster and support the data collected through other means.

This study undertook a documentary review technique for secondary data collection in an effort to collect additional information to augment the questionnaire results and interview results and so to provide a more significant interpretation of the data gathered. Studies that are now available frequently differ in terms of study design, operational quality, and study subjects. According to Saunders et al. (2012), the document review method supports the opinions or claims made in academic writing and may also highlight some difficulties that have gone unnoticed by other methods. How they approach the research question could vary, by increasing the validity of evidence collected.

Data Analysis

As noted earlier, following the quantitative data collection, data analysis was performed using PLS-SEM with Smart PLS 4. According to Hair et al. (2022), the reason the researcher chose PLS-SEM is that the analysis relates to testing the theoretical framework of predictive perspective and the structural model comprises many observed variables and latent variables from which the research is required to perform exploration of factor structure before actual testing of the hypothesis. In this study, the hypotheses were designed using latent variables such as tourism performance and digital marketing metaverse, and their respective observed variables. Having the nature of these two kinds of variables, observed and unobserved variables, in the conceptual framework of this study, PLS-SEM was considered suitable for analyzing this kind of model.

Ethical Consideration

During the study, the researcher ensured participant confidentiality about the information collected, and all data were analyzed in a group without personal identity. The researcher discussed with participants the intention of the study before the research study was conducted. The researcher secured a research permit to introduce himself and request support from the participants to agree to participate in the study. All information cited was acknowledged in the references list.

Findings

Validation of the Research Model

After the data collection process, it was necessary to evaluate the quality of the model developed that is to validate the correlation between the data collected and the conceptual model developed. This was due to the fact that the conceptual model was developed using variables borrowed from various empirical evidence without data. It was necessary to ensure that the data collected were in line with the conceptual model. However, ensuring the quality of data through the validity and reliability process depends on the research methods used. In this study, PLS-SEM was used using Smart PLS 4 software. Therefore, the quality of research in terms of validity and reliability was ensured based on the PLS-SEM criterion during reflective measurement and structural model formulation as follows.

Analysis of Reflective Measurement Model

A reflective measurement model was formulated and run to analyze the validity and reliability of the study model. The validity of the measurement model was evaluated using the following criteria: indicators loadings, convergent validity and discriminant validity, and internal consistency to check if they all align with the recommended value established by previous scholars (Hair et al., 2022). The first

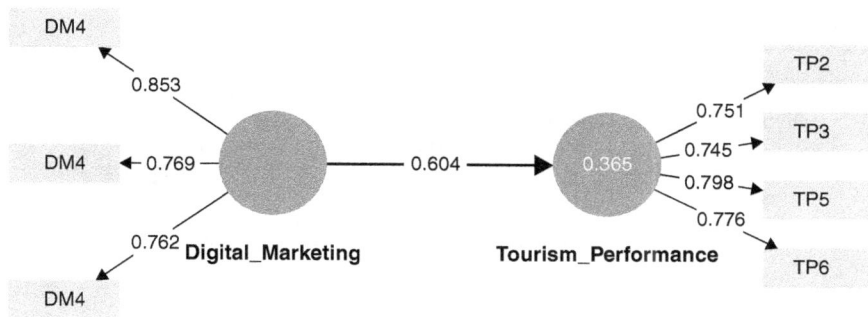

Figure 10.2 Reflective measurement model.

run of the reflective measurement model indicated the poor performance of the model in terms of validity and reliability of model constructs, namely, digital marketing metaverse and tourism performance restoration.

The model did not perform at the first run, due to the following indicator variables DM2 – "Digital communication" and DM5 – Digital showcase; "TP1: Increases in tourist visits and TP4 – Revenue generation" having a low loading of less than 0.7. It is argued that if the indicator scores a loading of less than 0.7, it affects the model performance due to the fact that in turn it will affect the value of average variance extracted (AVE), heterotrait–monotrait (HTMT) ratio of correlations, and composite reliability estimation. After the removal of these indicator variables, namely, DMS2, DM5, TP1, and TP4, the model performed well as demonstrated in the tables presented. Figure 10.2 presents the model constructs and their underlining indicator variables with their loading score which relate with the recommended loadings by Hair et al. (2017).

Table 10.1 is a table showing keys for all variables presented in Figure 10.2.

The following outputs were produced and presented to assess if the measurement model in Figure 10.2 has produced the scores that reflect the recommended score pattern suggested in previous studies for a model to be valid.

Loading

Hair et al. (2017) argued that a valid reflective measurement model is recognized when it produces a loading of 0.7 and above for all indicator variables in the model. Figure 10.2 and Table 10.2 show that all indicator variables have scored a loading of > 0.7 which is recommended in previous studies for a model to be valid.

Reliability and Convergent Validity

Cronbach's alpha and composite reliability were applied in assessing the internal consistency of the research instrument used during data collection. Scholars have

Table 10.1 Constructs in reflective measurement model

S/N	Constructs	Indicator variables
1	Digital marketing metaverse	DM1:Digital Information, DM3:Digital knowledge sharing, and DM4: Digital promotion
2	Tourism performance restoration	TP3:Increase visits by tourists TP2: Increased Market Share, TP5:Increased Product Purchase TP6:Increases in Foreign Currency

Table 10.2 Outer loading

Indicators/score	Digital marketing	Tourism performance restoration
DM1:Digital information	0.853	
DM3:Digital knowledge sharing	0.769	
DM4:Digital promotion	0.762	
TP3:Increase visits by tourists		0.751
TP2:Increased market share		0.745
TP5:Increase product purchase		0.798
TP6:Increase in foreign currency		0.776

recommended that a reliable model must produce both Cronbach's and composite reliability with a significant p-value > 0.7. In Table 10.3, all the constructs were found with a significant p-value >0.7 for both Cronbach's and composite reliability. This means that the current study model has reliable internal consistency aligned with the recommendation made by previous scholars for the reliability of the model. In a related perspective, AVE was used to assess the convergence validity in this study. For a model to achieve the degree required for convergence validity, it must produce AVE equal to or above 0.5. In Table 10.3, the results for AVE indicate that all the constructs have scored the value of AVE >0.5, which is recommended and accepted by previous scholars for the model to achieve convergence validity.

Divergence Validity

Divergence validity must be established to confirm that the construct in the measurement model is distinct from other constructs. Scholars have affirmed the use of the Fornell–Larcker criterion and the HTMT ratio of correlations in assessing the level of divergence validity. The Fornell–Larcker criterion proposed by Fornell and Larcker (1981) recommended that the square root of AVE in each latent variable can establish divergence validity if its value is larger than other correlation values among the latent variables. Table 10.4 was produced from the reflective measurement model in which the square root of the AVE was estimated by using the smart PLS 4 software and written in the diagonal of the table. The results in Table 10.4 suggest that the square root of AVE in each latent variable value is bigger than other

Table 10.3 Reliability and convergence

	Cronbach's alpha	Composite reliability (rho_a)	Composite reliability (rho_c)	Average variance extracted (AVE)
Digital marketing	0.709	0.710	0.838	0.633
Tourism performance	0.773	0.789	0.851	0.589

Table 10.4 Fornell–Larcker criterion

	Digital marketing	Restoration of performance in tourism business
Digital_Marketing	**0.796**	
Restoration of performance in tourism	0.604	**0.768**

Table 10.5 Divergence validity

Construct relationship	HTMT ratio
Restoration of performance in tourism – i.e., digital marketing	0.782

correlation values among the latent variables. Hence, based on the Fornell-Larcker criterion, this study achieved the recommended value for divergence validity.

Further analysis of the divergence validity was done using HTMT criteria, scholars have recommended that the reflective measurement model should produce an HTMT value of less than 0.8 (Hair et al., 2022). Since the maximum value produced in this study is 0.782 below the 0.85 threshold (i.e., the most conservative HTMT value), divergence validity is established in the model (vide Table 10.5).

Evaluation of the Structural Model

The evaluation of the structural model is based on four criteria, namely, collinearity assessment, path coefficient assessment, model explanatory power, and predictive power. In this assessment, R square (F-Size), the variance inflation factor (VIF), and the p-value were used. The structural model is presented in Figure 10.3.

Model Explanatory Power and Predictive Power

The R square represents the existing variance in each independent variable and is a measure of the explanatory power of the model, also referred to as predictive power. The assessment of explanatory power and predictive power of a model with

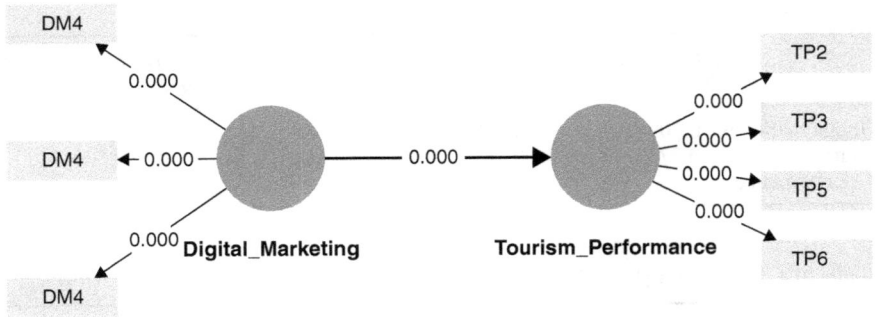

Figure 10.3 Structural model.

Table 10.6 R square

Independent variable	R-square	R-square adjusted
Performance of tourism business	0.365	0.363

R square ranges from 0 to 1 and is recognized. A higher value close to 1 indicates a greater explanatory power.

Table 10.6 indicates that the R square of the model is 0.365 which aligns with Hair et al. (2017), who recommended that for a model to predict, it must produce R^2 values ranging from 0 to 1. This means that the variance explained by the independent variable, namely, digital marketing metaverse ($R^2 = 0.365$) has a satisfactory predictive power.

Collinearity Analysis

In this study, collinearity in the structural model (Figure 10.3) is checked for potential collinearity problems in an effort to see if any of the variables should be eliminated, merged, or transformed into a higher-order latent variable. For a model construct to suffer from a collinearity problem, it should produce a variance inflation factor above 5 (Hair et al., 2022). Table 10.7 indicates that no constructs suffered from the collinearity problem since their VIFs were lower than 5.

Path Analysis and Hypothesis Testing

The study was designed and planned to analyze the contribution of digital marketing metaverse in restoring the performance of tourism business after the crisis of COVID-19. The researcher assessed the hypothetical relationship between the predictor variable (digital marketing metaverse) and the outcome variable (restoration of business performance in tourism). The structural model was run to assess

Table 10.7 Collinearity statistics

Indicator variables	VIF
DM1	1.699
DM3	1.468
DM4	1.289
TP2	1.621
TP3	1.675
TP5	1.483
TP6	1.400

Table 10.8 Path coefficients (Mean, STDE, T value, and p-values)

	Original sample (O)	Sample mean (M)	Standard deviation (STDEV)	T statistics (|O/STDEV|)	P-values
Digital marketing – performance of tourism business	0.604	0.607	0.038	16.022	0.000

the hypothesis, and the results are presented in Figure 10.3, which indicate that the path coefficient of both the hypothetical relationship of the independent variable to the dependent variable indicates a significant level using the p-value of less than 0.05.

Furthermore, Figure 10.3 shows the influence of each indicator variable which has scored a p-value of less than 0.05. Hair et al. (2022) recommended a p-value of 0.5 or less for a model hypothesis to be significant. Hence, in this study, all the indicator variables showed a significant contribution to the influence of independent variables on identified dependent variables.

Table 10.8 shows the path coefficient of the predictor variable (digital marketing metaverse) on the outcome variable (restoration of performance in tourism business). This is well predicted using the p-value at less than or equal to 0.05. Hence in this study, the influence of digital marketing metaverse on the restored performance of tourism business is significantly important.

Discussion

This study hypothesized on the existence of digital marketing metaverse impacting the restoration of performance of tourism business after COVID-19 and matching the business crisis in sub-Saharan countries. The finding of this study has revealed a significant influence of digital marketing metaverse for the restoration of tourism performance after the COVID-19 business crisis in sub-Saharan countries. The finding from quantitative data has been supported by the finding from qualitative data. Respondent 1 opined that *"Digital marketing metaverse is practically a combination of various and unique technologies which help to attract the attention of customers' behaviors to ward tourism destination even before their visit.* Similarly, Respondent 4 advocated that *"Metaverse marketing makes it easy for tourists to find the information they need, information on the goods they need to purchase, and all this is through adherence to existing traditions and cultural practices of the relevant tourist destination."* Furthermore, Respondent 3 said that *"It support virtual tourism, it allow a person to visit tourism destination virtually regardless of distance. In this case lockdown tourism destination does not affect metaverse tourism."*

These findings are in tandem with the observation made by Sánchez-Amboage et al. (2023), who observed and argued that digital marketing makes it easier to connect state management agencies with tourism businesses and customers, bringing benefits to the state and businesses in many aspects. The study supports the resource-based view theory, as described by Barney (1991), who argues for the importance of resources in promoting the performance of firms and industries. This implies that when tourism enterprises adopt digital marketing metaverse resources, they easily transform to meet their performance expectations and standards which in turn lead to improved performance of the tourism sector through access to critical market information and bridging the existing technological knowledge gaps.

This study supports the findings from prior studies (Velentza & Metaxas, 2023; Ndekwa, 2015; Bayelegne et al., 2022). Notably, Ndekwa (2015) argued that the adoption of information and communication technologies (ICTs) across the globe has altered the norms of the game and expectations of the new modes of economic activities in which ICT has promoted the performance of different sectors. On the other hand, Bayelegne et al. (2022) advocated that digital marketing metaverse has no boundaries and tourism sector can use any of the relevant devices such as smartphones, tablets, laptops, televisions, games, digital billboards, and media such as social media, search engine optimization, videos, content, email, and a lot more to promote access and enjoyments of tourist destinations. These findings imply that digital metaverse is a solution in restoring the performance of tourism and other businesses

Moreover, the findings and experiences imply that the findings relating to the adoption of digital marketing metaverse and tourism performance can be transferred and applied to other contexts altogether to evidence the importance of digitalization of tourism through metaverse. This means that according to the findings of the study, metaverse technologies are not affected by contextual variables.

While the findings of this study are in congruence with prior studies that support the influence of digital marketing metaverse on tourism performance, the explanation of how digital marketing restores business performance differs from one study to another. Notably, the examples given in the current study provide an explanation of the significant influence of digital marketing metaverse based on such attributes as digital information, digital knowledge sharing, and digital promotion. In the interview conducted, this observation was supported by Respondent 2 who asserted as follows:

> In tourism everything starts with would be tourists searching the internet for online information about the aspired tourist destination as a basis for stimulated further decision-making by prospective tourists through tourist information sharing via the virtual world.

From a different perspective, Razouk (2023) found that digital marketing metaverse channels such as social media and online information by travel agencies tend to restore tourism performance through fostering engagement and ensuring the visibility of the tourist destination. In his other studies, Ndekwa (2021a) found and concluded that mobile marketing metaverse significantly enhances tourism performance through facilitated online payment, online advertisement, and online promotion all of which, in turn, increase market share, sales, and overall revenue of the business.

In a related perspective using different attributes, Bhandari and Sin (2023) provided that digital marketing can reach and engage with their audience and drive business growth significantly. They added that digital marketing hotels can raise their business standing and revenue and encourage clients and customer attraction and retention as well as customer loyalty. In agricultural tourism, Ndekwa et al. (2023) found that marketing digitization can and does support digital advertisement, communication, and promotion and allow easy payment methods. This implies that digital marketing metaverse is a contextual variable in which its impact differs from one context to another. Furthermore, the finding implies that the contribution of digital metaverse differs from one personal standing to the other persons and from one business to another form of business.

Conclusion and Implications

This study established and concluded that digital marketing metaverse is significant in the restoration of performance of tourism and implicitly other business sectors as well. The possible explanation is based on the fact that metaverse offers digital information, digital knowledge sharing, and digital promotion which in turn promote increases in tourist visitations, market shares, product purchases, and national access to foreign currency. This study concludes that the contribution of digital metaverse differs from one personal viewpoint to another even in the same context and from one business to the other forms of business. This implies that metaverse technology is not a generic innovation but rather a flexible intervention in handling customer needs in the tourism sector.

This study concluded that digital marketing metaverse is key to restoring the performance of business in the tourism sector. The findings imply that digital metaverse is a solution to the business crisis by restoring the performance of tourism including its applicability in other contexts and sectors. The findings also imply that one context on the digital marketing metaverse can be transferred to and applied in other contexts of businesses.

Recommendation for Further Studies

This study was done in Tanzania, which is one of the countries in sub-Saharan Africa. It is recommended that future studies be conducted in other areas of sub-Sahara Africa to provide more evidence on the influence and contribution of metaverse in restoring the performance of businesses including and beyond tourism.

References

Adel, M. A. (2023). The role of Metaverse to create an interactive experience for tourists. *Journal of Association of Arab Universities for Tourism and Hospitality, 24*(1), 242–269.

Barney, J. B. (1991). Firm resources and sustained competitive advantage. *Journal of Management, 17*, 99–120.

Bhandari, R., & Sin, M. V. A. (2023). Optimizing digital marketing in hospitality industries. *Startupreneur Bisnis Digital (SABDA Journal), 2*(1).

Buhalis, D., Lin, M. S., & Leung, D. (2023). Metaverse as a driver for customer experience and value co-creation: Implications for hospitality and tourism management and marketing. *International Journal of Contemporary Hospitality Management, 35*(2), 701–716. https://doi.org/10.1108/IJCHM-05-2022-0631

Chaniago, A., & Arief, M. (2022, July 26–28). *Digital marketing and destination management models in shaping tourist behaviour*. Proceedings of the 5th European International Conference on Industrial Engineering and Operations Management, Rome, Italy.

Creswell, J. W. (2014). *Research design: Qualitative, quantitative and mixed methods approaches* (4th ed.). Sage.

Desai, V. (2019). Digital marketing: A review. *International Journal of Trend in Scientific Research and Development*. Special Issue Fostering Innovation, Integration and Inclusion Through Interdisciplinary Practices in Management, 196–200.

Fornell, C. G., & Larcker, D. F. (1981). Evaluating structural equation models with unobservable variables and measurement error. *Journal of Marketing Research, 18*(1), 39–50.

García, C., & Navarrete, M. C. (2022). Vaccine tourism: A new entrepreneurship. *Journal of Tourism and Heritage Research, 5*(4), 1–10.

Gnanapala, W. K. A., & Sandaruwani, J. A. R. C. (2016). Socio-economic impacts of tourism development and their implications on local communities. *International Journal of Economics and Business Administration, 2*(5), 59–67.

Gvaramadze, A. (2022). Digital technologies and social media in tourism. *European Scientific Journal, 18*(10), 28. https://doi.org/10.19044/esj.2022.v18n10p28

Hair, J. F., Babin, B. J., & Krey, N. (2017). Covariance-based structural equation modeling in the journal of advertising: Review and recommendations. *Journal Advert, 46*, 163–177. https://doi.org/10.1080/00913367.2017.1281777

Hair, J. F., Hult, G. T. M., Ringle, C. M., & Sarstedt, M. (2022). *A primer on partial least squares structural equation modeling (PLS-SEM)*. Sage.

Hettiarachchi, H. (2019). *The impact of the tourism industry on socio-economic development of informal and small scale business community: A case study from Ella divisional secretariat division of Badulla district*. Proceedings of the 6th International Conference on Multidisciplinary Approaches (iCMA). Faculty of Graduate Studies, University of Sri Jayewardenepura. SSRN. https://ssrn.com/abstract=3497396

Israilow, A. M. (2021). Importance of digital marketing and digital marketing strategies and tools in hospitality industry and tourism. *"Iqtisodiyot va innovatsion texnologiyalar" ilmiy elektron jurnali, 4*(52), 283–289.

Kannan, P. K., & Li, H. (2017). Digital marketing: A framework, review and research agenda. *International Journal of Research in Marketing*, *34*(1), 22–45.

Kavenuke, R., Matimbwa, H., Samwel, L., Jummane, H., Kapinga, E., & Ndekwa, A. G. (2017, August 10–11). *Mobile money payment adoption in tourism: Incidence from SMES from Zanzibar*. 17th International Conference on African Entrepreneurship & Small Business. University of Dar es Salaam Business School, Dar es Salaam Tanzania.

Lincoln, Z. (2013). Socio-economic and cultural impacts of tourism in Bangladesh. *European Scientific Journal*, *2*, 326–331.

Nassima, B., & Insaf, D. (2023). E-tourism in Algeria: New digital marketing strategy. *The International Journal of Business Management and Technology*, *7*(4), 24–41.

Ndekwa, A. G. (2014). Factors influencing adoption of information and communication technology (ICT) among small and medium scale enterprises (SMEs) in Tanzania. *International Journal of Research in Management and Technology*, *4*(5), 273–280.

Ndekwa, A. G. (2015). Determinants of adopter and non adopter of computerizing accounting system (CAS) among small and medium enterprises (SMEs) in Tanzania. *International Journal of Innovative Science, Engineering & Technology*, *2*(1), 438–449.

Ndekwa, A. G. (2017). *Factors influencing adoption of mobile money services among small and medium enterprises in Tanzania tourism sector* [PhD thesis, The Open University of Tanzania].

Ndekwa, A. G. (2021a). Connecting mobile marketing events and business performance during Covid-19 business crisis in Africa. *International Journal of Marketing & Human Resource Research*, *2*(4), 178–188.

Ndekwa, A. G. (2021b). COVID 19 and higher education institution operation: Unpacking the potential of higher education through acceptance of online learning. *International Journal of Education, Teaching, and Social Sciences*, *1*(1), 48–58. https://doi.org/10.47747/ijets.v1i1.491

Ndekwa, A. G. (2023). Linking business owner's market capability and mobile marketing adoption: Experience from Tanzania. In M. Valeri (Ed.), *Family business in tourism and hospitality*. Tourism, Hospitality & Event Management. Springer. https://doi.org/10.1007/978-3-031-28053-5_6

Ndekwa, A. G., & Katunzi, M. T. (2016). Small and medium tourist enterprises and social media adoption: Empirical evidence from Tanzania tourism sector. *International Journal of Business and Management*, *11*(4), 71–80.

Odhiambo, N. M. (2021). Tourism development and poverty alleviation in sub-Saharan African countries: An empirical investigation. *Development Studies Research*, *8*(1), 396–406. https://doi.org/10.1080/21665095.2021.2007782

Ponde, S., & Jain, A. (2019). Digital marketing: Concepts and aspects. *International Journal of Advanced Research*, *7*(2), 260–266. http://dx.doi.org/10.21474/IJAR01/8483

Rahardja, U. (2022). Social media analysis as a marketing strategy in online marketing business. *Startupreneur Bisnis Digital (SABDA Journal)*, *1*(2), 176–182.

Ramukumba, T. (2019). Analysing tourism stakeholder's perceptions of the attractiveness and competitiveness of the garden route as a tourist destination. *African Journal of Hospitality, Tourism and Leisure*, *8*(1), 1–9.

Razouk, C. B. (2023). The effect of digital marketing on tourism sector during Covid-19: An empirical study for Morocco. *International Journal of Management and Commerce Innovations*, *11*(1), 265–277. https://doi.org/10.5281/zenodo.8119885

Sánchez-Amboage, E., Membiela-Pollán, M. E., Martínez-Fernández, V., & Molinillo, S. (2023). Tourism marketing in a metaverse context: The new reality of European museums on meta. *Museum Management and Curatorship*, *38*(4), 468–489. https://doi.org/10.1080/09647775.2023.2209841

Saunders, M., Lewis, P., & Thornhill, A. (2012). *Research methods for business students* (6th ed.). Harlow Pearson Educational Limited.

Velentza, A., & Metaxas, T. (2023). The role of digital marketing in tourism businesses: An empirical investigation in Greece. *Businesses*, *3*, 272–292. https://doi.org/10.3390/businesses3020018

Wibawa, P. A., Astawa, P., & Sukmawati, M. R. (2022). Digital marketing and sustainable tourism for tourist villages in Bangli regency. *International Journal of Glocal Tourism*, *3*(2), 89–99.

Wonbera, T. W. (2019). The socio-economic and environmental impact of tourism industry on people of Arba Minch and its surroundings. *Journal of Tourism, Hospitality and Sports*, *40*, 1–5. https://doi.org/10.7176/JTHS

Zakarneh, B., Annamalai, N., Alquqa, E. K., Mohamed, K. M., & Salhi, N. R. (2024). Virtual reality and alternate realities in Neal Stephenson's "Snow Crash". *World Journal of English Language*, *14*(2), 244–252

11 Virtual Tourism as the Sustainable Future of Travel

Igor Mavrin and Corina Tursie

Introduction – Tourism in the New Era

The world has entered its traveling era after World War II, with constant growth in international travel (Statista, 2023), with the exception of the COVID-19 pandemic period, with a strong fall in 2020 and 2021, and the start of the recovery in 2022. As can be seen in Figure 11.1, international travel started to grow in the 1950s, with more significant rates during the 1970s, 1980s, and 1990s. The new millennium only accelerated the trends, with the main reasons found in the rise of the flight industry through low-cost agencies and affordable traveling (Papatheodorou, 2021; Valeri, 2023, 2024). However, the 2020 COVID-19 pandemic reversed the trend, at first completely stopping tourism, and after that continuing the process at a slower pace. Post-pandemic travel, starting in 2022, brought the new global phenomenon, revenge tourism (Singh et al., 2022; Kumar & Seth, 2022; Then & Yulius, 2022), with travelers overcrowding the airports with international travel intentions, but still not reaching the pre-pandemic levels of international travel. However, recent numbers (UNWTO, 2023a) show that the international travel market could reach the full pre-pandemic numbers in 2023, with the estimation of 80–95% of pre-pandemic numbers.

Another phenomenon was also emerging, induced by COVID-19 pandemic travel restrictions – the virtual tourism.

Virtual Transitions of Tourism Processes

Tourism is changing more quickly than expected, not only in the context of the COVID-19 pandemic but also related to the concepts of sustainability and culpability, especially regarding ecologically responsible forms of travel, with increasing awareness about the negative effects of all forms of traffic, with emphasis on air travel. Not all views on the COVID-19 situation are completely negative, although there were implications that the year 2020 was the beginning of the end of globalization (Forbes, 2021), the process that was already endangered before the pandemic (The Economist, 2020). There are views that "the temporary processes of de-globalisation are giving the global tourism industry a unique chance for a re-boot – an unrepeatable opportunity to re-develop in

DOI: 10.4324/9781003497004-13

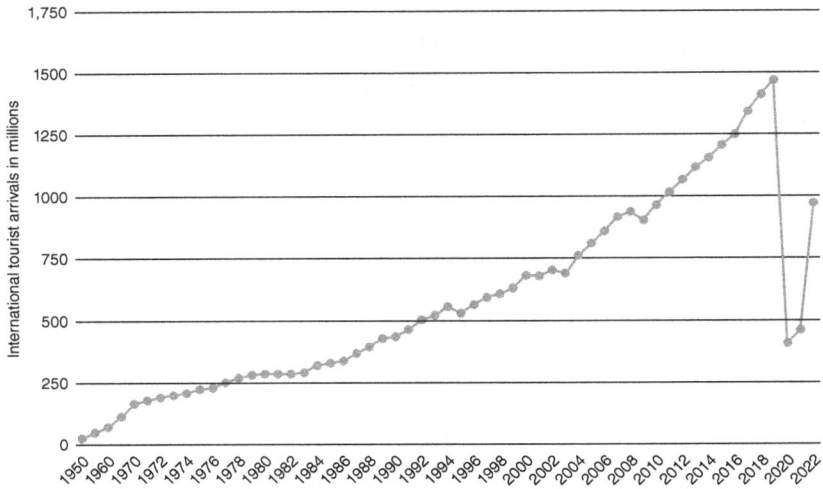

Figure 11.1 Number of international tourist arrivals worldwide from 1950 to 2022.
Source: Statista, 2023.

line with the tenets of sustainability" (Niewiadomski, 2020, p. 654), and also opinions that the pandemic is a wake-up call for the tourism industry, which could lead toward compassionate tourism (Stankov et al., 2020). Hussain (2021) implies a "holistic normal" concept of tourism after the pandemic, or a regenerative tourism concept – as opposed to the "new normal" and the return to the old normal. The holistic normal tourism concept means that destinations and visitors are a part of the living system embedded in the natural environment and that they act under nature's rules and principles. Initiatives worldwide are concentrated on policy-focused solutions for a global restart of tourism after the pandemic, based on the sustainability trend. UNCTAD (2020) suggested transformation in five priority areas: mitigation of socio-economic impacts on livelihoods, boosting competitiveness and building resilience, innovation and digital transformation of the tourism sector, fostering sustainability and green growth, and coordinating and creating partnerships to restart and transform the sector in accordance with the Global Sustainability Goals. The global digital shift encouraged by the pandemic will inevitably lead to the implementation of new technologies in the tourism and travel sector. Travelers are already heavily relying on technology, like online and mobile platforms, but also social media, searching for travel inspiration and recommendations, with the emergence and rise of virtual tourism (WTTC, 2020). The concepts of immersive tourism and virtual tourism have emerged in the past 20 years, allowing travelers to explore destinations, either in situ, enhanced by technology, or distantly, without leaving their homes or places of living.

Literature Review: Immersive Realities and Tourism

Immersive technologies have been the focus of scientists' interest since the 1990s, which coincides with their intensive emergence. Milgram and Kishino (1994) introduce the "virtuality continuum" concept, with the real environment on one end of the continuum and the virtual environment on the other (Figure 11.2). Real environments consist of only real objects, while virtual ones contain only virtual objects. Mixed reality (MR) environments are the ones in which real-world and virtual-world objects coexist and are presented together in a single display.

Three immersive technology concepts could be considered significant for tourism: AR, where virtual information is overlaid over the real world, VR, where real-life experience is designed in the virtual environment, and mixed reality (MR), a technology that provides the possibility for the coexistence of both virtual and real worlds (K S & Sinnoor, 2020). The three concepts, VR, AR, and MR, could be united within one term – immersive reality (IR), made from different features and functions enabling interaction with the virtual world (Ercan, 2020). The most recognizable, most commonly used, and most appropriate concept of IR for the times of the pandemic and the travel ban is VR because it provides a traveling experience without the need for an actual trip. Immersivity as a concept denotes the inherent quality of objects in general, and also of mediated and delineated, both real and imagined spaces, and it could be monitored interdisciplinary and multidisciplinary, with the incidence of immersive spaces in theme parks, films, theater, video games, and learning environments (Freitag et al., 2020). Virtual reality is the focus of interest for many scientific fields and disciplines. Psychology, as one of those disciplines, deals with the issues of presence, in this case, mediated by information technology, that provides the feeling of being in an external world (Riva & Waterworth, 2013). VR has numerous social and psychological characteristics, such as not only the ability to simulate activity and disembodied identity but also anonymity, identity expansion, deliberate impersonality, the ability to have multiple virtual personalities, and so on (Puchkova et al., 2017). All of the social and psychological characteristics of VR technology represent a potential for creating specialized tourism products for specific niches, creating and expanding the market for virtual tourism.

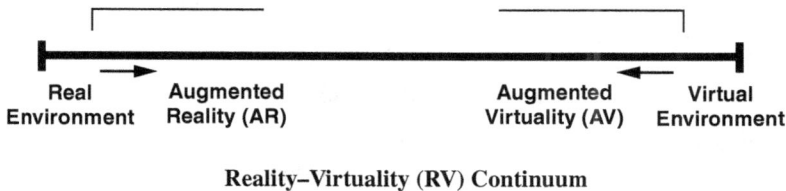

| Real Environment | Augmented Reality (AR) | Augmented Virtuality (AV) | Virtual Environment |

Reality–Virtuality (RV) Continuum

Figure 11.2 Milgram and Kishino's mixed reality on the reality–virtuality continuum (Milgram & Kishino, 1994).

Immersive Technologies in Tourism Research Trends

Since 2000, there has been a significant focus of researchers' interest on the themes of implementation of technology in tourism and cultural heritage. Mosaker (2001) has been researching possibilities of VR reconstruction for historical sites. There has also been research on web-based virtual tours for the purposes of destination marketing as the new trend in the early 2000s (Cho et al., 2002). 3D concepts were also a part of the researchers' focus: the use of 3D video games in museums (Lepouras & Vassilakis, 2005), e-tourism through 3D visualization based on 3D game-like applications (Berger et al., 2007), and also the use of 3D virtual environments for tourism (Huang et al., 2013).

From 2000 to 2010, researchers have mainly been focusing on 3D models and VR, while the focus from 2011 to 2020 has shifted to AR and other emerging immersive technologies. VR technology could significantly impact the future of family tourism (Whittington, 2014). VR usage in tourism could be considered a paradigm shift, with VR potentially substituting real travel with virtual travel (Taufer & Ferreira, 2019). However, the implementation of VR as a substitute for tourism could face some issues related to the authenticity of the tourist experience (Guttentag, 2010). AR could also provide an innovative way to enhance visitor experiences and preserve heritage (tom Dieck & Jung, 2017; Tussyadiah et al., 2018).

VR, AR, and emerging immersive technologies are especially important for the future of cultural tourism. Futuristic projections for the period until 2095 suggest that the cultural offer and the related tourism offer will be dominated by both AR and VR technologies (Richards, 2020). In that context, it is also important to mention the developing concepts of immersive heritage tourism (Bec et al., 2019) and immersive cultural tourism (Mavrin et al., 2020), with both terms suggesting that immersive technologies and videogames will impact the future of the heritage experience and interpretation, and the future of cultural tourism. 3D modeling and reality caption could be exploited in heritage preservation and in immersive visitor experiences used for tourism consumption (Little et al., 2019). Immersive technologies have already entered the tourism scene. "Virtual tours of museums, cultural heritage sites, and 360 live videos of events like music concerts are already allowing travellers to visit far off places and participate in events without leaving their homes" (Hawkinson & Klaphake, 2019, p. 66). Immersive technologies have great potential for tourism marketing, allowing a preview of the destination before the trip, but it could also be used as tourism enhancement at the location, and also as a travel substitute.

Immersive Technologies and Travel – Positive Versus Negative Sides

While immersive tourism implies a wider range of technological concepts (virtual reality, augmented and mixed reality, serious videogames, and other immersive technologies), leaving the option for travelers to actually visit the location or experience it from a distance, mediated by technology, the focus of this chapter will be set on virtual tourism conception. Virtual tourism could be defined as experiencing the destination without the need to travel (Stainton, 2020). Virtual space could be seen as tourist space (Pilarska & Tomczykowska, 2018) and it has a primary potential to attract disabled people, but it also targets the broader public interested in experiencing protected heritage areas (Voronkova, 2018). Key types of virtual tourism could be summarized in the following way: experiencing the destination before the actual travel, visiting destinations without leaving your home, visiting places of the past, visiting inaccessible areas, and visiting non-existing areas (Stainton, 2020). Until recently, the use of VR at the global level was predominantly characteristic for the pre-travel stage (GlobalData, 2019). VR could be used as a powerful tool for destination marketing, as research has shown that VR evokes more emotions in customers than brochures (Driescher et al., 2017), it helps people to generate more positive emotions toward a destination (Griffin et al., 2017), and it has the potential to engage tourists by enhancing their behavioral intentions, creating a need to visit destinations (Jung et al., 2017). The advantages of virtual tourism could also be summarized in terms of cost-effectiveness, health and overall safety, time savings, physical disability friendliness, low technological demands, new employment opportunities, and the emergence of new technologies (Sirnisava Rao & Krantz, 2020). Immersive technologies and even video games based on real locations could also be monitored as travel supplements (Mavrin et al., 2022), with both the potential to promote destinations and create new, virtual tourism niches.

Virtual travel is based on the combination of hardware and software solutions developed to stimulate one or more of the existing senses to create perceptually-real sensations (EITC):

- "Vision: (3D display, Fulldome, Head-mounted display, Holography).
- Auditory: (3D audio effect, High-resolution audio, Surround sound).
- Tactile: (Haptic technology)
- Olfaction: (Machine olfaction)
- Gustation: (Artificial flavor)."

The synesthetic experience could enhance the perception of reality in virtual environments and provide the platform for the future of virtual tourism. However, immersive tourism and virtual tourism could also have potentially negative impacts. Apart from depriving the destinations of their tourism income, some other issues could emerge, in terms of interpretation, health, security, ethics, and legality. Some authors doubt that online materials could provide an adequate experience for tourists who have decided to make a physical journey (Pilarska & Tomczykowska, 2018). Other potentially negative impacts include the divided brain, possible security issues emerging from the lack of attention paid to real-world activities overlaid with immersive content, digital addiction, breakdown of truth, and political polarization and radicalization, but also legal issues, such as privacy and data collection, payment and compensation, copyrights and intellectual property, cybercrime, public/private issues, and user agreements (Hawkinson & Klaphake, 2019). Other negative sides of virtual tourism incorporate the lack of real experience, the dependence of poorer countries on tourism revenues, health risks from immersion, limited reminiscence, and social implications (Rahman & Bhowal, 2017).

The emergence of the COVID-19 pandemic induced the production of research focused on the possibilities of virtual tourism as a substitute for physical travel. Research with the direct link to COVID-19 and VR as a substitution for real travel shows that there is an intention of tourists to use VR technology to travel virtually during and after a crisis like the outbreak of the COVID-19 pandemic (Sarkady et al., 2021), while in some earlier pre-pandemic research (Italy4Real, 2017) there was little or no interest for virtual travel. Virtual reality was also monitored as a tool for building resilience in small and medium tourist enterprises during the COVID-19 pandemic (Sánchez & Palos-Sanchez, 2021).

The global digital shift induced by the COVID-19 pandemic will inevitably have a lasting impact on all human activities, including tourism. Fear of the disease has temporarily repulsed a part of the tourism market from traveling. Even after the end of the pandemic, without the health issues, the global terrorism threat overshadows the tourism industry, making room for innovative, technology-based travel concepts. Constant advances in immersive technologies will make them less expensive and more accessible for wider audiences, while travel will become more expensive, due to security risks and sustainability trends focused on lowering carbon-dioxide emissions.

Never-Ending Tourism Concept, Artificial Intelligence, and the Metaverse Shift of Travel

One of the technology-driven trends in tourism is the never-ending tourism. The expression was coined by the Digital Innovation in Tourism Observatory (2022), and it refers to the possibility of extending the tourist experience in time and space, both before and after the trip, with the possibility of exploiting digital content in various directions, such as inspiring users and facilitating access to services, enriching onsite experience, and proposing contents and goods online. Exploring the never-ending tourism scenarios until 2030, Prodinger and Neuhofer (2023) provide three potential developments: the presence of technology in all areas of life, including travel; the preserved physical travel as an escapism from the omnipresent technology; and the scenario in which virtual spaces and digital twins dominate the everyday life, including tourism.

Artificial intelligence (AI) had a strong influence on global societies at the beginning of 2023, with functional chatbots entering the Internet mainstream. The AI concept is also part of the development in travel, tourism, and hospitality. AI systems could be both stand-alone systems, and also embedded within the existing systems and applications, including "recommender systems, personalization systems and techniques, conversational systems (chatbots and voice assistants), forecasting tools, autonomous agents, language translation applications, and smart tourism destinations" (Bulchand-Gidumal, 2022, p. 1950). Other authors connect AI with value co-creation in tourism, with different technologies influencing the process, that is, service robots, self-service kiosks enabled by AI, machine learning and natural language processing, chatbots, and metaverse tourism (Solakis et al., 2022). Although there are fears that AI could become a threat to labor in the tourism sector, Kirtil and Askun (2021, p. 226) emphasize that "along with governmental policy support, AI systems can be designed to reduce the negative effects of automation in industry," resulting with sustainable development in tourism, and AI can also have a role in tourism startups (Filieri & Raguseo, 2021). Strong AI emergence in 2023 brings a new perspective on the potential influence on tourism, hospitality, and travel. Development of software for AI-driven image and video generation, headed by Midjourney, has the strong potential to produce virtual environments used for virtual tourism faster than would be possible only with human experts included in production. Quicker development of virtual environments would consequently lead to the greater offer of virtual tours, generating the larger audiences, viz., virtual travelers.

Tourism is also being strongly influenced and impacted by the Metaverse shift. The phenomenon can be defined as "expansive virtual space where users can interact with 3D digital objects and 3D virtual avatars of each other in a complex manner that mimics the real world" (EICT), but it can also mostly overlap with the definition of cyberspace (Ravenscraft, 2023). Buhalis et al. (2023) introduced the metaverse tourism ecosystem concept, with the metaverse co-creation space as the connection between the tourism industry, consumers/tourists, and community and peers, before, during, and after the travel. Tourism and hospitality industries could benefit from the concept, with the creation of metaverse experiences with lifestyle, promotions, designing amusements, and designing adventures (Gursoy et al., 2022). Metaverse as the virtual environment with avatars of real existing people could also become a new platform for developing a new niche of tourism, based on virtual travels, without the need to leave your home or the place of residence. Metaverse tourism could lead to new ways for the tourism and hospitality industry to build their more sustainable futures.

Virtual Travel Policy Framework

In an era in which digitalization has a disrupting force, impacting society in new and often hard-to-predict ways, the tourism sector was an early adopter of information and communication tools. Platforms, online payments, and social media greatly impacted the tourism sector. Prior to the pandemic, a World Bank's research (Salem & Twining-Ward, 2018) acknowledged the complete change of the tourism landscape in only ten years because of the emergence of digital platforms. Changes occurred in the way travel was traditionally researched, bought, sold, experienced, and shared, while user-generated content became an important and widely used source of travel information. After the COVID-19 pandemic, other transformations are in the making. AI chatbots allow people to receive tailor-made travel recommendations, based on specific instructions, such as who they are traveling with or what activities they would be interested in.

According to UNWTO, digital transformation in the tourism sector "will grow to include location-based services, artificial intelligence, augmented and virtual reality and blockchain technology" (UNWTO, 2023b, p. 40), transforming the tourism offer toward a more attractive, efficient, inclusive, and economically, socially, and environmentally sustainable one.

In order to achieve these transformations, UNWTO proposes an open innovation approach for the tourism ecosystem stakeholders – from governments to corporations, academia, startups, or entrepreneurs – aiming to accelerate technology implementation. UNWTO is prompting governments to adopt policy measures that could facilitate the digital transformation of tourism (UNWTO website, digital transformation) in terms of smart travel, smart destinations, and job creation. For the intended purposes, smart travel facilitation includes smart visas, borders, security processes, and appropriate infrastructure, while smart destinations refer to the process by which destinations leverage technology to be more sustainable, accessible, and inclusive. Job creation refers to a two-folded preoccupation: toward new employment opportunities that could compensate for the effects of job eradication due to automation, and toward boosting technology- and skills-based education.

Several priority areas were defined by UNWTO (2023b) regarding tourism as a vehicle for achieving SDGs. Digitalization is one of them and is primarily seen as an opportunity to enhance competitiveness in the tourism sector, by improving visitor experience. In this context, the policies and regulations needed to be developed refer to digital protection standards to promote data privacy and cybersecurity. Another policy preoccupation refers to developing fair competition standards for tourism businesses, keeping up with evolving digital technologies.

The European Union's (EU's) General Data Protection Regulation and the recently adopted Digital Services Act and Digital Marketing Act (European Commission website, the Digital Services Act package, 2022) are putting the EU at the forefront of creating safer online environments, by making digital companies – such as online travel and accommodation platforms, among others – accountable for the content posted on their platforms.

At the level of the EU, narratives regarding the EU's digital agenda were developed in the last years, in the context of the Europe 2020 Strategy. The COVID-19 pandemic was seen as an opportunity to restart tourism and to build a long-term strategy toward a sustainable recovery of the sector, in line with the EU's strategic priorities: economic resilience, green transition (the EU Green Deal), and the digital transition. The current policy program defining "Europe's Digital Decade," as the vision for Europe's digital transformation by 2030 (Decision (EU), 2022), assumes the digitalization of businesses as a key action area. The other important parts of the EU's digital compass involve digital skills, secure and sustainable digital infrastructure, and public services digitalization. The digital transition in tourism is meant to take into consideration the opportunities and challenges for different stakeholders: business, public sector, tourism managers, and visitors.

For business, digitalization is expected to produce innovative business models; hence, special attention is given to promoting incubators connecting tourism and tech companies. During the 2014–2020 multiannual financial framework (MFF), the COSME program aimed to support "the digital transformation of tourism entrepreneurs, particularly SMEs and start-ups . . . and boost innovation along the tourism value chain through capacity building, training, coaching, technical assistance, prototyping, business matchmaking, financial advice, awareness raising" (COSME, 2020). In the new MFF 2021–2027, the COSME program is being continued by the Single Market Program and Invest EU. It is out of projects such as "EU DigiTour" (COSME, 2020, Funded projects) that we could expect to see transnational change-making solutions, enabling digital technologies for the tourism sector: AR/VR, AI, IoT, and big data.

From the point of view of local authorities and tourism managers – digitalization supports destination management and smart tourism. Since 2019, the EU organizes a competition to award the title of European Capital of Smart Tourism, to cities exhibiting exemplary practices of innovative, smart, and inclusive measures implemented in tourism. Being "a digital tourism city" means offering innovative "information, products, services, spaces and experiences adapted to the needs of the consumers through ICT-based solutions and digital tools" (Smart Tourism Capital, 2023). The digitalization of local administrations responsible for tourism and increased capacity in mastering tourism data are regarded as necessary achievements to facilitate the smart management of tourist flows at the destination level. A smart tourism city would offer its visitors innovative digital solutions such as integrated contactless tourist services, from access to museums, to public transportation, digital access tools in hotels, or enhancing visitor experience through digital tools (VR).

For visitors, digitalization promises to offer tailor-made experiences, which can be enjoyed at any point of the tourism cycle: before, during, and after a visit. However, public policies aiming to increase digital skills are an important precondition. According to the European Commission's Digital Economy and Society Index – DESI (2022), socio-demographic factors influence the levels of digital skills: only 29% of the retired and inactive Europeans have at least basic digital skills; also, only 46% of those living in rural areas have at least basic digital skills compared to people living in urban areas (61%) (DESI, 2022, p. 23). Another limit to overcome through cohesion policies is represented by the fact that, due to limited infrastructure in developing countries, 46% of the world's population does not have access to the Internet (UNWTO, 2021).

Conclusion: Toward the Virtual Future of Travel

Although the classical form of travel shall not perish, it could be expected that immersive technologies would lead toward both touristic experiences enhanced by augmented or mixed reality, together with the emerging forms of immersive realities, and with the separate strand of tourism – the virtual travels, as the potential new market that will include current travelers, former travelers, and current non-travelers.

As Table 11.1 implies, AR and MR will enhance the touristic experiences on-site, with immersiveness as the intensifier of the travel adventure. In the case of VR technology, immersiveness is taking place in completely artificial environments,

Table 11.1 Immersive tourism typology – technological framework and tourism types

Type of immersive technology	Tourism type	Main characteristics
Augmented reality (AR)	On-site immersive tourism	Enhancing the touristic experience by augmenting the scenery with technological solutions.
Mixed reality (MR)	On-site immersive tourism	Blurring the difference between virtual and real environments and enhancing
Virtual reality (VR)	Virtual tourism	Travelers immersing into completely artificial environments, based upon real or imagined worlds.

both in the case of creations of new surroundings and re-creations of existing ones or the perished worlds.

Virtual Tourism – The Motives Behind the Future Development

The main case for the virtual strand of tourism or the virtual shift for the future of travel could be found within the following key motives:

- The post-pandemic travel mode – As the global international travel numbers show, tourism has partially recovered from the COVID-19-induced travel reductions, but the numbers are still far off the pre-pandemic levels. Part of the pre-pandemic travel market was not impacted by the revenge tourism rebound in 2022, and although there could be 80–95% of pre-pandemic travels in 2023, the case still stands for the new travel habits formed in the lockdown periods of 2020 and 2021 for part of the travelers.
- Security issues – Terrorism and war threats will also shape the future of virtual travel. Global insecurity that marked the first two decades of the 21st century was temporarily overshadowed by the health issues from the pandemics but will remain one of the main drivers of the global processes. Ongoing wars (i.e., the Russo-Ukrainian war started in February 2022; Israel–Palestine recent conflict started in October 2023) will shape part of the travel market, with the conflict areas left out from the touristic processes leaving the opportunity for the virtual substitute of travel, for both residents of conflict areas and the travelers with intentions to visit conflict areas. Other types of conflicts, without the use of arms, are also emerging, like the potential new cold war between East and West, which could also (temporarily) suspend international travels and stimulate virtual travels as a substitute form of tourism.
- Sustainability trends – The sustainability imperative has entered the official global policies, with the 17 Sustainable Development Goals as frontrunners, and the EU implementing sustainability within their crucial documents, leading to

the Brussels effect of sustainability in the 27 EU member states. Sustainability has also spilled into the travel and tourism industry, with the flight shame trend and sustainable tourism concept as imperative. Sustainable tourism means less travel, opening the opportunity to virtual substitutes of tourism.

- Aging societies – Constant extension of expected life span leads toward rising numbers of active senior citizens, silver surfers from the Baby Boom generation, open toward new experiences, including travel and digital transition. Although people in their 60s and 70s still possess enough mobility for active vacations and cultural tourism experiences, the physical requirements of tourism for people in their 80s lead to travel deprivation. Virtual forms of travel would provide a substitute for travel, including new explorations, and virtual revisiting of destinations from previous decades, motivated by nostalgia.
- Millennials and Generation Z digital nativity and travel orientation – Younger Millennials and GenZ are digital natives, open toward new technologies, often embracing new tech solutions and including them in their everyday lives. As those generations also travel much more than the previous generations during the youth age, virtual forms of travel could become an addition to their tourism habits.
- Economic trends and the rising tourism prices – Inflation, the rise of consumer prices starting in 2022, with the recession threat in 2023, and the constant rise of travel costs, could also result in new traveling trends, like choosing domestic travel over international, and the virtual travel as the more affordable choice.
- Technological advancements and AI – Virtual reality is becoming more realistic with technological advancements, and AI could further improve the creation of realistic virtual worlds based on real existing locations, the locations from the past, and the locations from imagination. The more realistic virtual worlds could stimulate virtual tourism trends in the future.

All of those trends will have a significant impact on shaping the landscape of virtual tourism in the following decades. Apart from the virtual travels that are possible only in the virtual worlds (visiting the past, visiting the imagined landscapes), that will emerge into a completely new form of tourism, involving both existing travelers and non-travelers, other sub-forms of virtual tourism (visiting unreachable areas, and visiting virtual replicas of existing locations) should become the real focus of policy makers and tourism developers. As it could become both the threat to existing locations thriving on touristic growth (i.e., developing countries dependent on touristic revenue), and the solution for the destinations burdened by overtourism, tourism policies and digital policies should provide solutions for the issues. The policies should at the same time be focused on stimulating the development of virtual travel products for endangered locations and compensating the potential revenues lost for the tourism-dependent locations. Another level of threat could also emerge from the potential exploitation of virtualization of existing, or lost historical sites by the private investors, and even new forms of (virtual) neo-colonialism and cultural appropriation. Global policies at the UNWTO level and global solutions for the potential exploitation should be implemented during the development phase, to protect thriving societies dependent on tourism revenues, and to prevent unwanted scenarios from museums, because repatriation of virtual legacies could become even harder to achieve than the one with physical objects. Virtual tourism could co-exist with the current tourism model, and even bring a more sustainable approach to the industry considered one of the biggest polluters in the world (mass tourism, CO_2 emission from flights, etc.). However, this is possible only if all the stakeholders prepare a sustainable legal and economic platform for the virtual future of travel.

References

Bec, A., Moyle, B., Timms, K., Schaffer, V., Skavronskaya, L., & Little, C. (2019). Management of immersive heritage tourism experiences: A conceptual model. *Tourism Management, 72*, 117–120.

Berger, H., Dittenbach, M., Merkl, D., Bogdanovych, A., Simoff, S., & Sierra, C. (2007). Opening new dimensions for e-tourism. *Virtual Reality, 11*(2–3), 75–87.

Buhalis, D., Leung, D., & Lin, M. (2023). Metaverse as a disruptive technology revolutionising tourism management and marketing. *Tourism Management, 97*, 104724. ISSN 0261-5177. https://doi.org/10.1016/j.tourman.2023.104724

Bulchand-Gidumal, J. (2022). Impact of artificial intelligence in travel, tourism, and hospitality. In Z. Xiang, M. Fuchs, U. Gretzel, & W. Höpken (Eds.), *Handbook of e-tourism*. Springer. https://doi.org/10.1007/978-3-030-48652-5_110

Cho, Y. H., Wang, Y., & Fesenmaier, D. R. (2002). Searching for experiences: The web-based virtual tour in tourism marketing. *Journal of Travel and Tourism Marketing, 12*(4), 1–17.

COSME – Programme for the Competitiveness of Enterprises and small and medium-sized enterprises. (2020, March 4). *Call "innovation uptake and digitalisation in the tourism sector"* (COS-TOURINN). https://ec.europa.eu/info/funding-tenders/opportunities/portal/screen/topic-details/tourinn-01-2020 and funded projects https://ec.europa.eu/info/funding-tenders/opportunities/portal/screen/how-to-participate/org-details/999999999/project/101038104/program/31059643/details

Decision (EU) 2022/2481. (2022). *Of the European Parliament and of the Council establishing the digital decade policy programme 2030*. OJ L 323, pp. 4–26. https://eur-lex.europa.eu/eli/dec/2022/2481/oj and https://digital-strategy.ec.europa.eu/en/policies/europes-digital-decade

Digital Economy and Society Index – DESI. (2022). *European commission, annual report*. https://ec.europa.eu/newsroom/dae/redirection/document/88764

Driescher, V., Lisnevska, A., Zvereva, D., Stavinska, A., & Relota, J. (2017). Virtual reality: An innovative sneak preview for destinations. *Virtual Reality in Tourism – Current News, Trends & Feature Articles*. www.virtual-reality-in-tourism.com/research/virtual-reality-innovative-sneak-preview-for-destinations/

The Economist. (2020). Has covid-19 killed globalisation? *The Economist*. www.economist.com/leaders/2020/05/14/has-covid-19-killed-globalisation

EITC. (n.d.). *Immersive technology (VR, AR, MR, XR, the metaverse)*. www.eitc.org/research-opportunities/new-media-and-new-digital-economy/computer-vision-immersive-technology-and-digital-content/immersive-technology/immersive-technology

Ercan, F. (2020). An examination on the use of immersive reality technologies in the travel and tourism industry. *Business and Management Studies: An International Journal, 8*(2), 2348–2383. https://doi.org/10.15295/bmij.v8i2.1510

European Commission website, The Digital Services Act package. (2022). https://digital-strategy.ec.europa.eu/en/policies/digital-services-act-package

Filieri, R., & Raguseo, E. (2021). Artificial intelligence (AI) for tourism: An European-based study on successful AI tourism start-ups. *International Journal of Contemporary Hospitality Management*, *33*(11), 4099–4125. Emerald Publishing Limited. https://doi.org/10.1108/IJCHM-02-2021-0220

Forbes. (2021). Does COVID-19 mean the end for globalization? *Forbes*. www.forbes.com/sites/imperialinsights/2021/01/08/does-covid-19-mean-the-end-for-globalization/?sh=4a7c9189671e

Freitag, F., Molter, C., Mücke, L. K., Rapp, H., Schlarb, D. B., Sommerlad, E., Spahr, C., & Zerhoch, D. (2020). Immersivity: An interdisciplinary approach to spaces of immersion. *Ambiances* [Online], Varia. http://journals.openedition.org/ambiances/3233

GlobalData. (2019). *Virtual reality in travel & tourism – thematic research* (Report). Research and Markets.

Griffin, T., Giberson, J., Lee, S. H. M., Guttentag, D., & Kandaurova, M. (2017). *Virtual reality and implications for destination marketing*. Travel and Tourism Research Association: Advancing Tourism Research Globally. TTRA Annual International Conference Proceedings. https://core.ac.uk/download/pdf/84289283.pdf

Gursoy, D., Malodia, S., & Dhir, A. (2022). The metaverse in the hospitality and tourism industry: An overview of current trends and future research directions. *Journal of Hospitality Marketing & Management*, *31*(5), 527–534. https://doi.org/10.1080/19368623.2022.2072504

Guttentag, D. A. (2010). Virtual reality: Applications and implications for tourism. *Tourism Management*, *31*(5), 637–651.

Hawkinson, E., & Klaphake, J. (2019, December 8–9). *The emergence of legal and ethical issues in immersive tourism in Asia* (pp. 66–71). Proceedings of 210th IASTEM International Conference, Nagoya, Japan.

Huang, Y. C., Backman, S. J., Backman, K. F., & Moore, D. (2013). Exploring user acceptance of 3D virtual worlds in travel and tourism marketing, *Tourism Management*, *36*, 490–501.

Hussain, A. (2021). A future of tourism industry: Conscious travel, destination recovery and regenerative tourism. *Journal of Sustainability and Resilience*, *1*(1), 1–10.

Italy4Real (2017). *Italy4Real*. https://italy4real.com/

Jung, T., tom Dieck, M. C., Moorehouse, N., & tom Dieck, D. (2017). *Tourists' experience of virtual reality applications*. IEEE International Conference on Consumer Electronics (ICCE). https://ieeexplore.ieee.org/xpl/conhome/7886213/proceeding

Kirtil, I. G., & Askun, V. (2021). Artificial intelligence in tourism: A review and bibliometrics research. *Advances in Hospitality and Tourism Research (AHTR): An International Journal of Akdeniz University Tourism Faculty*, *9*(1), 205–233. https://doi.org/10.30519/ahtr.801690

K S, G., & Sinnoor, G. P. (2020). *Management of tourism experiences using immersive technology. Tourism and hospitality: Theories and practices*. Bharti Publications.

Kumar, V., & Seth, K. (2022). Push and pull factors influencing travel desire and revenge tourism intention in India post Covid-19. In G. Parkash, S. Chhabra, & S. Thakur (Eds.), *Perspective of revenge tourism* (pp. 64–72). Bharti Publications. ISBN: 978-81-19079-15-5.

Lepouras, G., & Vassilakis, C. (2005). Virtual museums for all: Employing game technology for edutainment. *Virtual Reality*, *8*(2), 96–106.

Little, C., Bec, A., Moyle, B. D., & Patterson, D. (2019). Innovative methods for heritage tourism experiences: Creating windows into the past. *Journal of Heritage Tourism*, *15*(1), 1–13. https://doi.org/10.1080/1743873X.2018.1536709

Mavrin, I., Mesić, H., & Šebo, D. (2020). *Towards the new model of heritage management – potentials of ICT in interpretation and presentation of urban legacy* (pp. 894–911). IMR

2020, Interdisciplinary Management Research XVI; The Josip Juraj Strossmayer University of Osijek, Faculty of Economics in Osijek [Postgraduate Doctoral Study Program in Management, Hochschule Pforzheim University, Croatian Academy of Sciences and Arts].

Mavrin, I., Šebo, D., & Glavaš, J. (2022). Immersive cultural tourism in the context of COVID-19 pandemic: Global perspectives and local impacts. *Ekonomski pregled: Mjesečnik Hrvatskog društva ekonomista Zagreb, 73*(5), 739–767. https://doi. org/10.32910/ep.73.5.4

Milgram, P., & Kishino, F. (1994). A taxonomy of mixed reality visual displays. *IEICE Transactions on Information Systems, E77-D*(12).

Mosaker, L. (2001). Visualising historical knowledge using virtual reality technology. *Digital Creativity, 12*(1), 1–194. https://doi.org/10.1076/digc.12.1.15.10865

Niewiadomski, P. (2020). COVID-19: From temporary de-globalisation to a re-discovery of tourism? *Tourism Geographies, 22*(3), 651–656. https://doi.org/10.1080/14616688.2 020.1757749

Osservatorio Innovazione Digitale nel Turismo. (2022). *Neverending Tourism: Che cos'è e come sta crescendo grazie all'Innovazione Digitale.* https://blog.osservatori.net/it_it/ neverending-tourism-significato

Papatheodorou, A. (2021). A review of research into air transport and tourism: Launching the annals of tourism research curated collection on air transport and tourism. *Annals of Tourism Research, 87*, 103151. ISSN 0160-7383. https://doi.org/10.1016/j. annals.2021.103151

Pilarska, A. A., & Tomczykowska, P. (2018). Virtual tourism space of cities. *Journal of Modern Science, 3*(38), 317–333. https://doi.org/10.13166/jms/99215

Prodinger, B., & Neuhofer, B. (2023, January 18–20). Never-ending tourism: Tourism experience scenarios for 2030. In B. Ferrer-Rosell, D. Massimo, & K. Berezina (Eds.), *Information and communication technologies in tourism 2023 proceedings of the ENTER 2023 eTourism conference* (pp. 288–299). https://link.springer.com/ chapter/10.1007/978-3-031-25752-0_31

Puchkova, E. B., Sukhovershina, Y. V., & Temnova, L. V. (2017). A study of generation Z's involvement in virtual reality. *Psychology in Russia: State of the Art, 10*(4). https://doi. org/10.11621/pir.2017.0412

Rahman, S., & Bhowal, A. (2017). Virtual tourism and its prospects for Assam. *IOSR Journal of Humanities and Social Science (IOSR-JHSS), 22*(2), 91–97.

Ravenscraft, E. (2023). What is the metaverse, exactly? *Wired.* www.wired.com/story/ what-is-the-metaverse/

Richards, G. (2020). Culture and tourism: Natural partners or reluctant bedfellows? A perspective paper. *Tourism Review, 75*(1), 232–234.

Riva, G., & Waterworth, G. A. (2013). Being present in a virtual world. In M. Grimshaw (Ed.), *The Oxford handbook of virtuality* (Chapter 12). Oxford University Press.

Salem, T. M., & Twining-Ward, L. D. (2018). *The voice of travelers: Leveraging user-generated content for tourism development.* World Bank Group. http://documents.worldbank. org/curated/en/656581537536830430/The-Voice-of-Travelers-Leveraging-User-Generated-Content-for-Tourism-Development-2018

Sánchez, M. R., & Palos-Sanchez, P. (2021). Virtual reality as tool for resilient tourism companies. In J. V. de Carvalho, Á. Rocha, P. Liberato, & A. Peña (Eds.), *Advances in tourism, technology and systems* (vol. 208). ICOTTS 2020. Smart Innovation, Systems and Technologies. Springer. https://doi.org/10.1007/978-981-33-4256-9_4

Sarkady, D., Neuburger, L., & Egger, R. (2021). Virtual reality as a travel substitution tool during COVID-19. In W. Wörndl, C. Koo, & J. L. Stienmetz (Eds.), *Information and communication technologies in tourism* (pp. 452–463). Springer. https://doi. org/10.1007/978-3-030-65785-7_44

Singh, T., Goel, R., Baral, S. K., Sehdey, S. L., & Gupta, S. (2022). *Revenge tourism: Reviving hotel industry in 4.0 era* (pp. 71–76). 2022 International Conference on Recent Trends in Microelectronics, Automation, Computing and Communications Systems (ICMACC), Hyderabad, India. https://doi.org/10.1109/ICMACC54824.2022.10093462

Sirnisava Rao, A. S. R., & Krantz, S. G. (2020). Data science for virtual tourism: Using cutting-edge visualizations: Information geometry and conformal mapping. *Patterns, 1.* https://doi.org/10.1016/j.patter.2020.100067

Smart Tourism Capital, Initiative of the European Union. (2023). *Competition for the European capital of smart tourism 2024: Guide for applicants.* https://smart-tourism-capital.ec.europa.eu/system/files/2023-04/Capitals_Guide_for_Applicants_2024.pdf

Solakis, K., Katsoni, V., Mahmoud, A. B., & Grigoriu, N. (2022). Factors affecting value co-creation through artificial intelligence in tourism: A general literature review. *Journal of Tourism Features*, 1–15. Emerald Publishing Limited. ISSN 2055-5911. https://doi.org/10.1108/JTF-06-2021-0157

Stainton, H. (2020). Virtual tourism explained: What, why and where. *Tourism Teacher.* https://tourismteacher.com/virtual-tourism/#8-types-of-virtual-tourism.

Stankov, U., Filimonau, V., & Vujičić, M. D. (2020). A mindful shift: An opportunity for mindfulness-driven tourism in a post-pandemic world. *Tourism Geographies, 22*(3), 703–712. https://doi.org/10.1080/14616688.2020.1768432

Statista. (2024). *Number of international tourist arrivals worldwide from 1950 to 2023.* https://www.statista.com/statistics/209334/total-number-of-international-tourist-arrivals/

Taufer, L., & Ferreira, L. T. (2019). Realidade virtual no turismo: Entretenimento ou uma mudança de paradigma? *Rosa dos Ventos – Turismo e Hospitalidade, 11*(4), 908–921.

Then, J., & Yulius, K. G. (2022). Motivation and interest in traveling of young traveler during revenge tourism. *Global Research on Tourism and Advancement, 4*(2), 2022.

tom Dieck, M. C., & Jung, T. H. (2017). Value of augmented reality at cultural heritage sites: A stakeholder approach. *Journal of Destination Marketing and Management, 6*(2), 110–117.

Tussyadiah, I. P., Jung, T. H., & tom Dieck, M. C. (2018). Embodiment of wearable augmented reality technology in tourism experiences. *Journal of Travel Research, 57*(5). https://doi.org/10.1177%2F0047287517709090

UNCTAD. (2020). *Life after lockdown: Rebuilding tourism globally, sustainably.* https://unctad.org/news/life-after-lockdown-rebuilding-tourism-globally-sustainably

UNWTO. (2021). *UNWTO inclusive recovery guide – sociocultural impacts of Covid-19, issue 2: Cultural tourism.* World Tourism Organization. https://doi.org/10.18111/9789284422579

UNWTO. (2023a, May). *World tourism barometer.* World Tourism Organization. https://en.unwto-ap.org/news/worldtourismbarometer_may2023/

UNWTO. (2023b). *Goa roadmap for tourism as a vehicle for achieving the sustainable development goals.* World Tourism Organization. https://doi.org/10.18111/9789284424443

UNWTO. (n.d.). *UNWTO website, digital transformation.* www.unwto.org/digital-transformation.

Valeri, M. (2023). *Tourism innovation in the digital era: Big data, AI and technological transformation.* Emerald Publishing.

Valeri, M. (2024). *Innovation strategies and organizational culture in tourism: Concepts and case studies on knowledge sharing.* Routledge Publishing.

Voronkova, L. P. (2018). Virtual tourism: On the way to the digital economy. *IOP Conference Series: Materials Science and Engineering, 463*(4). https://iopscience.iop.org/article/10.1088/1757-899X/463/4/042096

Whittington, A. (2014). Family vacation 2050: Socially and technologically-driven scenarios of the future of family travel, recreation and tourism. *Tourism Recreation Research, 39*(3), 379–396.

World Trade & Tourism Council. (2020). *To recovery & beyond: The future of travel & tourism in the wake of COVID-19*. Oliver Wyman. www.oliverwyman.com/content/dam/oliver-wyman/v2/publications/2020/To_Recovery_and_Beyond-The_Future_of_Travel_and_Tourism_in_the_Wake_of_COVID-19.pdf

12 Exploring Immersive Tourism

Insights From Padova's Tourism Managers

Mohammed Reda Boukhecha, Fatima Zahra Fakir, and Silvia Rita Sedita

Introduction

Visitors of Generation Z, born in the digital age, are driving a transformation in the tourism sector. This generation has a strong desire for co-creating unique travel experiences and tends to be well-behaved, risk-averse, and embracing smart tourism innovations (Buhalis & Karatay, 2022). Therefore, the introduction of immersive experiences at tourism and cultural heritage sites is expected to play a crucial role in enabling Generation Z to co-create transformative journeys.

Looking forward to the future of cultural heritage tourism, it is not just a journey into the past but also an investment in the future. Moreover, preserving and celebrating our cultural treasures today ensures that future generations can connect with the rich tapestry of human history and craft their own transformative experiences (Harrison et al., 2020). In this context, the integration of technology and digital features is a transformative force. These innovations have fundamentally changed the way visitors interact with historical and cultural artifacts, leading to enriched experiences and improved understanding. Technology, including audio guides, applications, AR, and VR, holds great potential in raising awareness and captivating visitors (Allal-Chérif, 2022). Likewise, many historical sites have embraced innovative tools and methods to bring preservation into the digital age (Marra et al., 2021).

Moreover, the infusion of technology has rejuvenated ancient artifacts and historical sites. For instance, AR applications can overlay digital information onto real-world environments, allowing visitors to see, hear, and interact with the past in unprecedented ways. For Instance, "Rekrei," an AR app, lets users reconstruct damaged historical artifacts using crowdsourced photos and 3D printing (Kirkpatrick, 2015). Similarly, the "Smart History" app utilizes AR to offer interactive tours of historical sites with 3D models and audio commentary (Yang & Wang, 2021).

Cultural heritage tourism involves exploring a specific location's cultural and historical elements, artifacts, and traditions (Yin et al., 2022). In this context, Padova, also known as Padua, in northeastern Italy, stands out as an exemplary destination with a wealth of cultural and historical significance. One facet of Padova's cultural heritage is its architectural treasures, such as the ancient University of Padova, established in 1222 and known for its rich architectural heritage (Cecchini

DOI: 10.4324/9781003497004-14

et al., 2018). Even more, the Basilica of Saint Anthony of Padua, a revered pilgrimage destination, is another architectural marvel that draws visitors from around the world (Porto et al., 2012). These architectural wonders not only enhance the city's visual appeal but also serve as enduring witnesses to its deep historical significance.

Furthermore, Padova is actively involved in conserving and researching its cultural heritage. The University of Padova, in fact, takes proactive steps to study and document the impact of seismic events on cultural heritage structures (Porto et al., 2012; Hofer et al., 2016), demonstrating the city's commitment to cultural preservation.

Thus, Padova's cultural heritage plays a significant role in attracting tourists, and it is considered a smart and cultural destination due to its efforts to integrate digital technologies into its tourism industry while also promoting its cultural heritage (Fakir et al., 2023). In this regard, the aim of this study is to comprehensively examine the impact of immersive digital technologies and digital content on tourists' experiences and engagement at cultural heritage sites in Padova, within the context of a smart cultural destination, and to understand how these factors are interconnected with tourists' perceptions and decision-making processes. In pursuit of this goal, we engaged in interviews with tourism experts in Padova to better understand how to offer an immersive experience that aligns with tourists' preferences.

Tourist Experiences in Smart Destinations

The Influence of Technology on Tourist Behavior

The influence of technology on tourist behavior is a compelling focus of interest within the tourism industry (Jeong & Shin, 2019). Technology's sway over tourist behavior is profound, and the establishment of a robust and sustainable information technology infrastructure in companies operating in the tourism and hospitality industry is recognized as a pivotal success factor. When tourists make online bookings, for instance, their behavior often leans toward impulsive decisions. This inclination can be fueled by online word-of-mouth or peer reviews. which encompass aspects such as hotel ratings, user-friendliness, hotel website functionality, and considerations of security and privacy (Müller et al., 2018). Furthermore, technology-integrated experiences in tourism possess the capacity to significantly augment tourist satisfaction and encourage repeated visits (Jeong & Shin, 2019). Given the escalating competition in the tourism marketplace, businesses are increasingly delving into the realm of smart technology to furnish personalized experiences (Neuhofer et al., 2015). Nevertheless, the comprehension of the role of smart technology in tailoring personalized experiences remains relatively limited (Neuhofer et al., 2015).

The first area of inquiry pertains to the acceptance of navigation apps and the subsequent ramifications of smart tourism technology, and the intention to utilize these applications. Zeng et al. (2022) investigated the intention to use navigation apps and the moderating effects on visitors' intention. By adopting a survey-based approach, they collected data from tourists visiting Lijang, an ancient town in Yunnan (China), and discovered that there is a relationship between having lower spatial cognitive capacity and the reliance on navigation apps for wayfinding and navigation. Hence, this underscores the critical importance of accounting for the technological requirements of diverse user groups, including those with diminished spatial perception, when comprehending tourist behavior (Zeng et al., 2022).

The second area of inquiry pertains to the pivotal role of electronic word of mouth (eWOM) via social media platforms in steering tourists' perceptions of a destination. Jalilvand and Samiei (2012) found support of eWOM being an influential factor in shaping tourists' perceptions, attitudes, and behaviors. Within this context, shared travel experiences on social media platforms emerge as a form of electronic e-WOM with the potential to wield significant influence over tourists' decision-making processes (Liu et al., 2018). From another point of view, eWOM can also shape individual identity. Berger and Greenspan Weed (2008) investigated adventure tourism, elucidating how technology shapes tourist identities through two distinct mechanisms. On one hand, it engenders new identities through novel experiences; on the other hand, it reinforces pre-existing ones via self-expression and validation. Importantly, the impact of technology on tourist identities is not uniform, varying depending on individual preferences, motivations, and social contexts. These evidences underscore the necessity for a nuanced comprehension of how technology influences tourist identities within the sphere of adventure tourism. In addition, eWOM can be the virtual place where a destination personality is built. Papadimitriou et al. (2013) delved into the impact of destination personality and affective image on tourists' behavioral intentions within the domain of domestic urban tourism. This study, grounded in surveys conducted with tourists visiting a domestic urban destination, illuminated the significant influence of destination personality and affective image in shaping the overall image formation of a destination. This, in turn, molds tourists' perceptions and evaluations, subsequently impacting behavioral intentions such as revisitation and the dissemination of positive word-of-mouth (Papadimitriou et al., 2013). To sum up, social media platforms offer travelers a virtual place where to share their experiences, concurrently presenting the identities of destinations, which can profoundly influence the perceptions and motivations of others to travel (Ba & Song, 2022). These platforms have emerged as critical sources of travel information, with tourists heavily relying on them to gather information about destinations, accommodations, and activities (Xiang & Gretzel, 2010). The act of tourists sharing their experiences on social media positively regulates their own tourism experiences, fostering heightened satisfaction and greater expectations for future trips (Ba & Song, 2022).

The third area of inquiry is about the use of technologies to improve the shopping experience of tourists. Al-Sulaiti (2022) explored the impact of technology-enabled facilities within mega shopping malls on tourists' behavior and revisit intentions. Employing the stimulus-organism-response (SOR) theory, the study elucidated the intricate relationship between technology-enabled facilities, destination image, tourists' behavior, and revisit intentions. The findings illuminated that advanced technology and innovative features significantly enhance the shopping experience, subsequently influencing tourists' perceptions and behaviors. This study imparts valuable implications for small management and marketing strategies, highlighting the pivotal role of technology in molding tourists' perceptions and behaviors. Furthermore, following the SOR model, Chen et al. (2022) ventured into the relationship between tourism experiences, the emotion of "fun," recommendations, and revisit intentions among Chinese outbound tourists. This research, employing structural equation modeling, discerned that memorable tourism experiences wield considerable influence over the emotion of "fun," subsequently shaping tourists' recommendations and their intentions to revisit. Such insights can be instrumental in the endeavors of destination marketers and managers, as they seek to tailor experiences that resonate with the preferences of Chinese outbound tourists (Chen et al., 2022).

Visitor Engagement in Smart Destinations

Engaging visitors is necessary for the prosperity and sustainability of smart destinations. Ivars-Baidal et al. (2017) emphasized that tourist involvement greatly influences the success of smart tourism initiatives, which are increasingly prevalent in European tourism, born out of smart city projects (Gretzel et al., 2015). These initiatives harness modern information and communication technologies (ICTs) to strengthen the connections between businesses, destinations, and tourists (Femenia-Serra et al., 2018). Destinations aspiring to become or enhance their status as smart tourism hubs must embrace a comprehensive approach to destination management. This involves formulating specific strategies and implementing efficient governance models, delineating data flows and essential datasets, fostering strategies for enhancing the environmental sustainability of the destinations, integrating crucial technologies and infrastructures, and developing skills and policies tailored for entrepreneurs and businesses within the tourism ecosystem (Figure 12.1).

A key factor impacting visitor engagement is the utilization of smart tourism technologies. Such technologies significantly bolster tourist satisfaction and their loyalty to a destination. They create memorable tourism experiences that heighten overall tourist satisfaction (Azis et al., 2020). Additionally, smart tourism technologies alleviate concerns and uncertainties related to trip planning and visits, fostering a sense of novelty (Goo et al., 2022). As Sustacha et al. (2023) highlight, smart destinations employ advanced technology and mobile digital tools to enhance visitor engagement, creating value for both tourists and destinations.

Nonetheless, visitor engagement in smart destinations is influenced by various factors. The quality of the visitor's relationship with the destination is one such element. Research has shown that visitor engagement positively influences environmentally responsible behavior, with this impact mediated by relationship quality (Zhou et al., 2020). Moreover, sensory experiences at destinations have been found to affect visitors' digital engagement with these places (Ai et al., 2022). Positive sensory experiences can lead to more favorable attitudes and loyalty toward the destination (Agapito et al., 2017). It is also evident that visitor engagement is a crucial factor influencing satisfaction with heritage destinations, as it mediates the

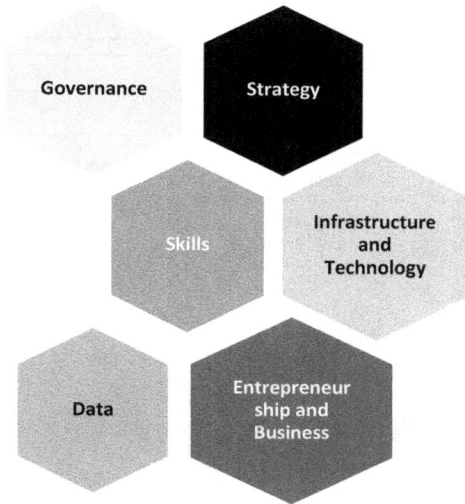

Figure 12.1 Key aspects for a smart destination development toolkit (EU guide on data for tourism destinations, 2022).

relationship between travel motivation, visitor experience, and the heritage desti-nation image (Sun et al., 2021). In the context of smart tourism ecosystems, the willingness of tourists to adopt smart practices and actively engage with various stakeholders is pivotal for the success of smart destinations (Femenia-Serra et al., 2018). This underscores the significance of visitor engagement as a driving force behind the development and implementation of smart tourism initiatives. Hence, a successful strategy for smart destinations to foster positive visitor engagement should be personalized, technology-driven, and aimed at enhancing the tourism experience by providing fast, accurate, and comprehensive information services to tourists (Lan et al., 2021). Consequently, the smart tourism program should prior-itize the practical use of smart travel measures in real-world applications, such as hotels and tourist sites, through sensor networks, the Internet of Things, and mobile Internet (Figure 12.2).

Thus, visitor engagement is a process characterized by frequent interactions between visitors and the destination, which serve to strengthen the psychological, emotional, and behavioral connections made by visitors (Zhou et al., 2020). Sig-nificantly, the acceptance of smart initiatives, whether by locals or visitors, can be a potential barrier to engagement, as observed in a study by Ye et al. (2021). In this case, it is imperative to adopt a comprehensive approach that takes into account the perspectives and experiences of both residents and tourists.

According to Santos-Júnior et al. (2020), smart tourism destinations harness the power of ICT to elevate the tourist experience and enhance the well-being of resi-dents. Their ultimate goal is to establish an ecosystem that prioritizes the needs of

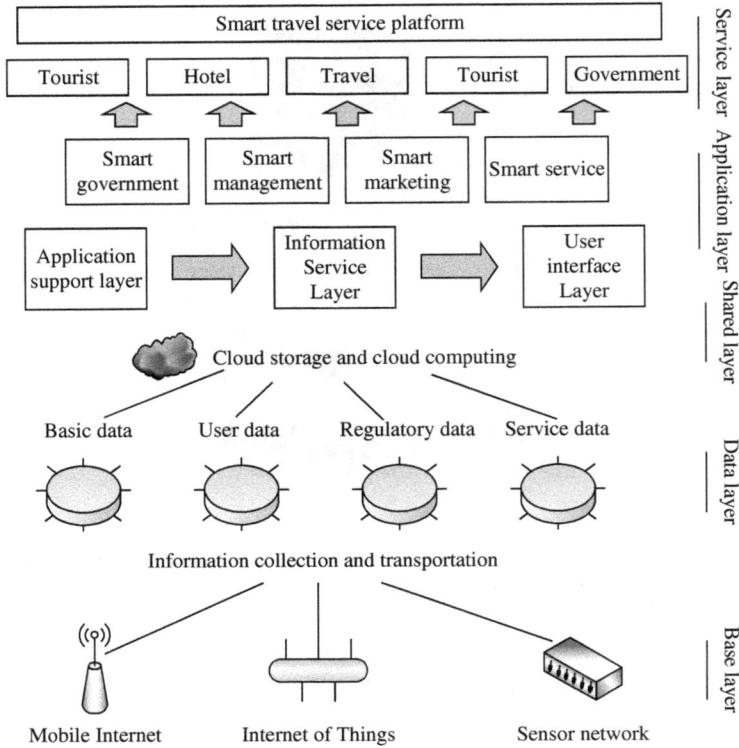

Figure 12.2 Tourism experience and construction of personalized smart tourism program under tourist psychology (Lan et al., 2021).

customers, offering highly personalized and automated experiences (Buhalis et al., 2019). Nevertheless, it is crucial to ensure that the development of smart destinations remains attentive to the welfare of both residents and tourists. One key aspect of addressing barriers is the emphasis on improving the quality of life for residents. Smart tourism cities, a subset of smart tourism destinations, place a stronger focus on resident well-being (Lee et al., 2020). By taking into account the preferences and requirements of residents, smart destinations can foster a more inclusive and sustainable environment. This can be achieved through the efficient utilization of technological, human, and social resources (Pencarelli, 2019). Additionally, it is of paramount importance to comprehend and address the concerns and challenges faced by tourists. Through the strategic deployment of technology, smart destinations can provide tourists with innovative and tailored experiences (Molinillo et al., 2019). Table 12.1 shows the World Best Smart Destinations nominees (World Travel Tech Awards, 2022).

Table 12.1 World Best Smart Destinations nominees (World Travel Tech Awards, 2022)

N°	Destination	The achievement
1	Abu Dhabi, UAE	The city has adopted various smart technologies to enhance the tourist experience, featuring a mobile application that delivers details on local attractions, events, and transportation.
2	Barcelona, Spain	Barcelona has integrated a range of smart technologies to enhance the well-being of both residents and tourists. These encompass smart lighting systems, waste management systems, and a mobile app offering real-time updates on public transportation.
3	Boston, USA	Boston has employed a variety of smart technologies to improve the tourist experience, utilizing a mobile app that offers information about local attractions, events, and transportation.
4	Copenhagen, Denmark	Copenhagen has implemented a suite of smart technologies to enhance the tourist experience, including a mobile application that provides insights into local attractions, events, and transportation.
5	Dubai, UAE	Dubai has embraced an array of smart technologies to enrich the tourist experience, featuring an application that offers real-time details about local attractions and events.
6	Finland, Sweden	Finland has adopted multiple smart technologies to enhance the tourist experience, incorporating an application that provides real-time information about local attractions and events.
7	Glasgow, Scotland	Glasgow has implemented a range of smart technologies to enhance the quality of life for both residents and tourists. These encompass smart lighting systems, waste management systems, and a mobile application that delivers real-time information on public transportation.

Smart Tourism Applications and Mobile Applications

Smart tourism is a multifaceted concept influenced by a range of elements, including the Internet, social media, smart devices, mobile apps, online games, augmented reality, virtual reality, blockchain, and cryptocurrency. The integration of smart technologies such as artificial intelligence, cloud computing, the Internet of Things (IoT), and mobile communication has empowered tourism destinations to enhance tourists' experiences (Jeong & Shin, 2019). These smart technologies allow for personalization, context awareness, and real-time monitoring, which elevate the smart tourism experience (Gretzel et al., 2015). Moreover, they foster better connectivity and information sharing, creating a networked environment for tourism stakeholders (Zvaigzne et al., 2023).

Stakeholders, in turn, play a vital role in the development and implementation of smart tourism technologies as they contribute to the creation, management, and delivery of intelligent tourism services and experiences (Gretzel et al., 2015). Additionally, the adoption of 5G technology further facilitates the advancement of smart tourism by improving the quality of tourism services and aligning with public needs (Liang et al., 2021)

Mobile applications represent a fundamental element of smart tourism, offering tourists convenient access to information and services. As a result, they have gained increasing influence within the tourism industry, shaping travel behaviors and enriching tourist experiences (Gupta et al., 2018). According to a survey conducted by eMarketer, travel-based mobile apps rank as the seventh most frequently downloaded app category, with nearly 60% of smartphone users regularly using travel apps during trip planning. Therefore, the role of mobile applications in smart tourism has been widely examined. Stakeholders view mobile apps, websites, QR code applications, and physical infrastructure as indispensable components of smart tourism technology (Hamid et al., 2023). In essence, mobile technologies and applications play a pivotal role in smart tourism by providing convenience, time saving, and cost reduction for consumers (Xu et al., 2019). The widespread adoption of mobile technology has established mobile communication as a significant channel for tourism organizations (Kim et al., 2008).

In parallel, tourism consumers are increasingly reliant on mobile apps for sourcing tourism information (Figure 12.3), prompting numerous studies exploring their usage patterns (Choi & Yoo, 2021). These applications often offer features such as

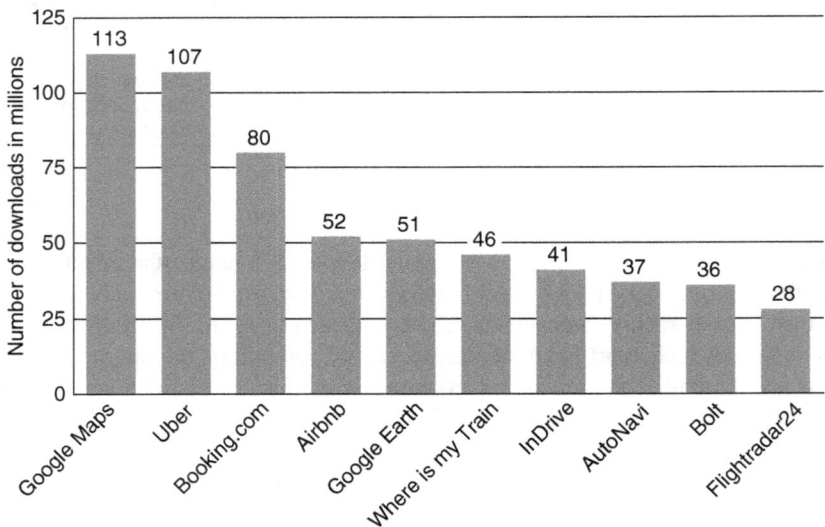

Figure 12.3 Most downloaded travel apps worldwide in 2022 (Statista, 2023).

interactive maps, real-time suggestions, personalized itineraries, and mobile payment options (Hamid et al., 2023). However, the development of mobile travel apps is critical for building trust and satisfaction among tourists and enhancing their engagement with tourism destinations (Imtiaz & Suki, 2022). Smartphone applications have been found to influence the information-seeking behavior of foreign independent travelers, with tourism-related apps being particularly favored (Lin et al., 2020). According to a report from the McKinsey Global Institute in 2018, numerous digital applications address practical and human concerns in smart cities. These applications have the potential to enhance key quality-of-life indicators by 10–30%, resulting in saved lives, reduced crime incidents, shorter commutes, a lighter health burden, and lowered carbon emissions.

Methodology

Within this section, we offer an overview of the interview results and conduct a thorough analysis. These interviews were conducted by experienced tourism professionals in Padova, providing a multifaceted perspective on the potential influence of metaverse technologies on the tourism sector. The interviews delved into several key areas:

- *The Influence of Metaverse and Innovation on the Tourism Experience.*
- *Aspects of Authenticity, Sustainability, and Local Engagement.*
- *Exploring Age and Generational Variances in Metaverse Adoption.*
- *The Effects of Leadership and Management on Metaverse Implementation.*

Qualitative Findings

Museums Tour Guide Padua

- **Immersive Technologies and AI's Role in Guiding**: The guide acknowledges the potential of digital technologies such as metaverse and AI in the tourism industry. She believes that AI will not replace human guides but can assist them by providing information faster. She states, "could AI will take my place as a guide? I say no, it will help me to do things faster."
- **AI as a Referral Tool**: The guide sees AI as a way to refer tourists to guides. She mentions, "you can ask chat GPT tourist guide in Barcelona, and he will give you a list, So AI can boost my business make me more visible."
- **Importance of Human Contact**: She emphasizes the importance of human contact in tourism and how local guides have more insight into the city, highlighting the irreplaceable role of a human guide in providing a dynamic and immersive experience.
- **Digital Features and Social Media in Guided Tours**: The guide discusses how she uses technology, like radios with amplifiers, to enhance her tours and ensure that participants can hear her clearly. She also uses social media for sharing photos and engaging with her clients.

- **Benefits of Digital Features**: She believes that digital features can provide a deeper understanding and interactive learning for tourists. For example, audio guides in museums and interactive elements like videos can enhance the tourist experience.
- **Preservation of Authenticity**: The guide emphasizes the need to curate accurate information when incorporating digital features, ensuring that the information presented is true to the historical and cultural context. She states, "you have to know exactly to say and be aware of your knowledge."
- **Challenges With Technology**: Tourist unfamiliarity with technology is a challenge she faces, particularly with older tourists who may struggle with using their phones or scanning QR codes.
- **Feedback From Tourists**: The guide receives feedback from tourists, with common suggestions for improvement. She cites an issue with audio guides being available only in Italian at the museum where she works, which hinders the experience for non-Italian-speaking tourists.
- **Target Tourist Profile**: Her target tourists, especially in the area, are German, and she suggests offering content in German to cater to their needs and potentially increase revenue.

These findings from the meeting with "Museums tour guide Padua" suggest that while technology, including metaverse and AI and digital features, can enhance the tourist experience, human guides are invaluable for providing local insights and ensuring the accuracy and authenticity of the information presented. Additionally, adapting to the preferences and languages of target tourists is essential for improving the overall experience and potentially increasing revenue.

Licensed City Tour Guide of Padua

- **Enhancing Visitor Experience**: The tour guide believes that digitalization and innovation in touristic cultural sites can enhance the overall visitor experience. She agrees that these technological advancements can improve the way tourists interact with and learn about cultural heritage.
- **Benefits to the Local Community and Economy**: She sees the benefits of digitalization and innovation extending to the local community and economy by improving educational experiences for individuals. While she does not elaborate further, it suggests that technology can have a positive ripple effect.
- **Sustainability and Environmental Respect**: The guide emphasizes the importance of sustainability and respect for the environment in digitizing touristic cultural sites. She suggests promoting eco-friendly transportation options for visitors and using renewable energy sources to power digital installations to ensure that the digitization is environmentally friendly.
- **Resident Involvement in Tourism Promotion**: She believes that residents can play a significant role in promoting Padova as a tourist destination. They can actively contribute to marketing and enhancing the city's appeal to tourists.

- **Methods for Resident Contribution**: She outlines various methods through which residents can contribute, such as sharing positive experiences on social media and travel platforms, providing local recommendations and tips, participating in community events, supporting local businesses, and engaging in volunteer activities related to tourism promotion.
- **Digitalization and Authenticity**: The guide acknowledges the potential risk of excessive digitalization harming the authenticity of cultural heritage sites in Padova. She believes that decisions regarding digitalization should be taken rationally and consider all aspects, including the preservation of authenticity.

These findings suggest that the Licensed tour guide of Padua is supportive of digitalization and innovation in the tourism sector but stresses the importance of balancing technological advancements with sustainability, environmental concerns, and the preservation of cultural authenticity. She also recognizes the vital role that residents can play in promoting and enriching the tourist experience in Padova.

Tourism Manager of the Veneto Region

Destination Management Approach: The Manager's approach to tourism is primarily a destination management approach, focusing on organizing, governing, and developing tourism in a holistic manner rather than merely marketing and communication.

Readiness for Digital Integration: He suggests that Padova is well-prepared for the integration of virtual reality and other digital technologies, citing its strong potential driven by the presence of local universities with expertise in technology, human resources, and a real interest in developing and using new technologies to promote and support visits to Padova.

Digital Initiatives in Museums: The Manager highlights museums and attractions in Padova that have adopted digital approaches, such as the Museum for Medicine (MUSME) and the Museum of Science and Natural History. He acknowledges that there are gaps and variances in the adoption of digital technologies in different museums.

Local Community Acceptance: The Manager believes that the local community is generally receptive to the adoption of new technologies in tourism. While there might be variations based on age and generational differences, he suggests that this approach could facilitate visiting experiences for a broad range of visitors.

Role of Policies and European Funds: He points out that policies and funding from European programs are instrumental in supporting the adoption of digital technologies in tourism. These initiatives aim to incentivize and sustain the use of digital technologies for promotion and commercialization.

Destination Management Organization (DMO): The Manager recognizes the need for a well-organized DMO in Italy, including in Padova, as part of the wider efforts to evolve and streamline the tourism industry. He highlights that DMOs play a crucial role in managing and promoting destinations effectively.

Public–Private Partnership: He emphasizes the importance of a public–private partnership in tourism, highlighting the necessity for cooperation between public and private sectors to address issues such as sustainability and the growing phenomenon of short-term rentals affecting residents and students in places like Padova.

Veneto Tour Guides Association

- **Enhancing Visitor Experience**: The organization acknowledges that digitalization and innovation in touristic cultural sites can enhance the overall visitor experience, indicating a positive stance toward technological advancements in tourism.
- **Benefits to the Local Community and Economy**: They recognize the benefits of digitalization and innovation by stating that it improves educational experiences for individuals and creates new job opportunities for local residents. This demonstrates a holistic perspective on the economic and educational impact of such advancements.
- **Sustainability and Environmental Respect**: The organization highlights the importance of sustainability and environmental respect in the digitization of touristic cultural sites. They suggest measures such as promoting eco-friendly transportation options, using renewable energy sources, and designing energy-efficient digital displays and devices to ensure sustainability and environmental friendliness.
- **Resident Involvement in Tourism Promotion**: They agree that residents can significantly contribute to promoting Padova as a tourist destination, emphasizing the potential role of locals in marketing their city.
- **Methods for Resident Contribution**: The organization believes that residents can contribute in various ways, including sharing positive experiences on social media and travel platforms, as well as supporting and recommending local businesses to tourists. This suggests the importance of resident engagement in enhancing the tourist experience.
- **Digitalization and Authenticity**: Similar to previous interviewees, the organization agrees that excessive digitalization could harm the authenticity of cultural heritage sites in Padova. They advocate for making rational decisions that consider all aspects, particularly the preservation of cultural heritage authenticity.

Discussion

The qualitative insights derived from these interviews provide valuable perspectives on the incorporation of immersive technologies into Padova's tourism landscape. The first theme centers on the potential of metaverse technologies to enhance the overall tourism experience, as illustrated by the Museum tour guide, a tourist guide who envisions the metaverse and AI as tools to make destination information more accessible and transform tourist experiences into immersive journeys. The second theme underscores the importance of sustainability and local community involvement, as highlighted by Licensed tour guide of Padua and the Tour Guides Association, emphasizing the need for eco-friendly practices and community engagement to preserve cultural and environmental authenticity. The third theme addresses potential generational disparities in Technology adoption, with the Tourism Manager of the Veneto Region acknowledging that older tourists may be less familiar with technology, potentially leading to gaps in adoption. The fourth theme emphasizes the critical role of leadership and destination management, with the Tourism Manager of the Veneto Region underscoring the importance of DMOs and effective leadership in guiding destinations toward digital transformation. A recurring subtheme is the need to strike a balance between traditional and digital approaches, ensuring digital technologies and AI complement rather than replace existing tourism practices to accommodate tourists with varying preferences and technological familiarity. These findings also acknowledge challenges, such as technological unfamiliarity among older tourists and the imperative for better alignment between technology and sustainability.

Conclusion

Through qualitative interviews, we uncovered the potential of immersive technologies, such as the metaverse and AI, in enriching the visitor experience by providing accessible destination information and immersive tours. However, the interviews also highlighted challenges like technology unfamiliarity, particularly among older tourists, and emphasized the importance of authenticity, sustainability, and the role of the local community in tourism promotion. Moreover, there is a need for quality, user enjoyment, and balance between traditional and digital approaches. In light of the integrated findings, several key recommendations can be proposed to further improve and advance smart cultural tourism in Padova.

- **Digital Education and Training Programs:** Implement digital education and training programs for local residents, especially older generations, to increase their familiarity and comfort with technology. This can help bridge the generation gap in technology adoption.
- **Enhanced Digital Content:** Develop high-quality, multilingual digital content about cultural products in Padova. This content should be engaging, informative, and authentic to attract and satisfy tourists with diverse interests.
- **Sustainable Technology Integration:** Ensure that the integration of metaverse technologies and other digital features in cultural heritage sites aligns with sustainability and eco-friendly practices. Explore the use of renewable energy sources for technology installations.
- **Community Involvement:** Encourage residents to actively participate in promoting Padova as a tourist destination. This involvement can include sharing positive experiences on social media, offering local recommendations and tips, supporting local businesses, and engaging in volunteer activities related to tourism promotion.

Theoretical Implications

Our research contributes significantly to the theoretical framework of smart tourism management by employing conceptual approaches and interviews to investigate the influence of smart technologies on the visitor experience. It deepens our understanding of tourist consumer behavior within the smart tourism ecosystem, elucidating how tourist behavior is shaped by various factors such as socioeconomic backgrounds, cultural norms, and personal preferences. Furthermore, the integration of smart technologies such as artificial intelligence, cloud computing, the IoT, and mobile communication has empowered tourism destinations to enhance tourists' experiences (Jeong & Shin, 2019). These theoretical frameworks not only advance our comprehension of tourist motivations and behaviors but also offer guiding principles for the development of impactful strategies within the tourism industry. Moreover, insights into the extensive impact of digital technology on tourist behavior create opportunities for innovative strategies aimed at enriching visitor experiences and boosting engagement levels (Gretzel & Scarpino-Johns, 2018). The strategic integration of digital tools and platforms provides tourism providers with the chance to craft immersive and interactive experiences that captivate contemporary travelers. These experiences transcend traditional boundaries, offering dynamic encounters that deeply resonate with the evolving preferences of modern tourists.

Management Implications

This research underscores the intricate process involved in the advancement of smart tourism and smart destinations. While management entities acknowledge the advantages associated with such progress, numerous destinations are actively striving to achieve efficient implementation. Consequently, politicians are increasingly recognizing the imperative to capitalize on the economic potential offered by the intersection of ICTs with tourism. Furthermore, this research aids in pinpointing specific factors or attributes within the destination that merit heightened attention and prioritization in the management process. These factors may encompass initiatives such as digital education and training programs, community involvement, and the establishment of destination management organizations. By acknowledging and giving priority to these essential factors, destination managers can allocate resources, formulate strategies, and execute measures more efficiently to tackle challenges and seize opportunities, thereby bolstering the overall management and advancement of the destination.

Limitations and Future Studies

Our qualitative study primarily delved into the insights provided by professionals operating within the tourism sector, inadvertently overlooking a thorough examination of the viewpoints held by tourists themselves. Incorporating the perspectives of tourists could have offered a more holistic understanding of the dynamics at play. Furthermore, while the qualitative interviews were conducted with knowledgeable professionals, it is important to acknowledge that this group represents only a specific segment of the tourism industry. Their insights might not fully capture the diverse range of perspectives present within the broader tourism landscape, potentially limiting the breadth of our findings. Thus, to further expand our understanding of tourist behavior and preferences in Padova and continue enhancing the city's tourism sector, several avenues for future research merit exploration:

- Longitudinal Studies: Conduct longitudinal studies over an extended period to capture the dynamics of tourist behavior and perceptions throughout the year. This would help identify seasonal variations and the impact of events or festivals on tourist preferences.
- In-Depth Tourist Perspectives: Future research should focus on obtaining in-depth perspectives directly from tourists visiting Padova, to gain a deeper understanding of their preferences, needs, and challenges when it comes to the integration of digital features in their experiences.
- Cross-Cultural Studies: Explore how cultural backgrounds and origin countries influence tourist behaviors and perceptions in Padova. This research can provide valuable insights for tailoring experiences to specific cultural preferences.
- Comparative Analysis: Extend the study to include comparisons with other tourist destinations to identify common trends and unique characteristics of Padova.
- Comparing similar and dissimilar destinations can offer insights into the specific advantages and challenges in Padova's tourism industry.
- Digital Literacy and Accessibility: Investigate tourists' digital literacy levels and their access to technology. Understanding the digital divide among tourists and its impact on their experiences can inform strategies for making digital features more inclusive.
- Impact of Events and Promotions: Explore how events, promotions, and marketing campaigns influence tourist perceptions and behaviors. Analyzing the effectiveness of different promotional strategies can provide guidance for enhancing tourist experiences.
- Sustainability and Eco-Friendly Practices: Investigate the extent to which tourists prioritize sustainability and eco-friendliness when selecting destinations. This research can help in designing and promoting eco-friendly initiatives and practices within Padova's tourism sector.
- Technological Innovation: Examine emerging technologies and their potential impact on the tourism sector. Investigate how technologies such as augmented reality, virtual reality, and blockchain can be integrated into Padova's tourism experiences.
- Local Community Engagement: Research the involvement of the local community in tourism promotion and management. Assess the effectiveness of initiatives that encourage locals to contribute to tourists' experiences and perceptions.
- Public–Private Partnerships: Further explore the potential for public–private partnerships to address various challenges in the tourism sector. Investigate the

role of such collaborations in improving accommodation availability, safety, and overall tourist satisfaction.

- Comparative Survey Platforms: Use a variety of survey platforms beyond Google Forms to reduce digital bias and reach a broader range of respondents, ensuring more diverse insights into tourist perceptions and behaviors.
- Visitor Segmentation: Develop a deeper understanding of different visitor segments, such as family tourists, solo travelers, business travelers, and cultural enthusiasts, to tailor experiences more effectively to their specific needs and preferences.

Acknowledgments

This study is supported by MIUR (Ministry of Education, Universities and Research, Italy) through a project in the framework of the National Recovery and Resilience Plan (NRP). Any opinions, findings, conclusions, or recommendations expressed in this material are those of the authors and do not necessarily reflect the views of the MIUR.

References

Agapito, D., Pinto, P., & Mendes, J. (2017). Tourists' memories, sensory impressions and loyalty: In loco and post-visit study in Southwest Portugal. *Tourism Management, 58*, 108–118. https://doi.org/10.1016/j.tourman.2016.10.015

Ai, J., Liu, Y., Hu, Y., & Liu, Y. (2022). An investigation into the effects of destination sensory experiences at visitors' digital engagement: Empirical evidence from Sanya, China. *Frontiers in Psychology, 13*. https://doi.org/10.3389/fpsyg.2022.942078

Allal-Chérif, O. (2022). Intelligent cathedrals: Using augmented reality, virtual reality, and artificial intelligence to provide an intense cultural, historical, and religious visitor experience. *Technological Forecasting and Social Change, 178*, 121604.

Al-Sulaiti, I. (2022). Mega shopping malls technology-enabled facilities, destination image, tourists' behavior and revisit intentions: Implications of the Sor theory. *Frontiers in Environmental Science, 10*. https://doi.org/10.3389/fenvs.2022.965642

Azis, N., Amin, M., Syafruddin, S., & Aprilia, C. (2020). How smart tourism technologies affect tourist destination loyalty. *Journal of Hospitality and Tourism Technology, 11*(4), 603–625. https://doi.org/10.1108/jhtt-01-2020-0005

Ba, D., & Song, L. (2022). The impact of after-travel sharing on social media on tourism experience from the perspective of sharer: Analysis on grounded theory based on interview data. *Wireless Communications and Mobile Computing, 2022*, 1–9. https://doi.org/10.1155/2022/7202078

Berger, I. E., & Greenspan, I. (2008). High (on) technology: Producing tourist identities through technologized adventure. *Journal of Sport & Tourism, 13*(2), 89–114. https://doi.org/10.1080/14775080802170312

Buhalis, D., Harwood, T., Bogicevic, V., Viglia, G., Beldona, S., & Hofacker, C. F. (2019). Technological disruptions in services: Lessons from tourism and hospitality. *Journal of Service Management, 30*(4), 484–506. https://doi.org/10.1108/josm-12-2018-0398

Buhalis, D., & Karatay, N. (2022, January 11–14). Mixed reality (MR) for generation Z in cultural heritage tourism towards metaverse. In *Information and communication technologies in tourism 2022: Proceedings of the ENTER 2022 eTourism conference* (pp. 16–27). Springer International Publishing.

Cecchini, C., Cundari, M., Palma, V., & Panarotto, F. (2018). Data, models and visualization: Connected tools to enhance the fruition of the architectural heritage in the city of
Padova. *Graphic Imprints*, 633–646. https://doi.org/10.1007/978-3-319-93749-6_51

Chen, Z., Wu, J., Gan, W., & Qi, Z. (2022, December). Metaverse security and privacy:
An overview. In *2022 IEEE international conference on big data (big data)* (pp. 2950–
2959). IEEE.

Choi, J., & Yoo, D. (2021). The impacts of self-construal and perceived risk on technology readiness. *Journal of Theoretical and Applied Electronic Commerce Research*, *16*(5),
1584–1597. https://doi.org/10.3390/jtaer16050089

Fakir, F. Z., Sedita, S. R., Andrian, G., & Marchioro, S. (2023). Smart cultural Padua ecosystem (Ecosistema culturale intelligente di Padova). *Regional Studies and Local Development (RSLD) Journal*, *4*(4), 117.

Femenia-Serra, F., Neuhofer, B., & Baidal, J. (2018). Towards a conceptualisation of smart
tourists and their role within the smart destination scenario. *Service Industries Journal*,
39(2), 109–133. https://doi.org/10.1080/02642069.2018.1508458

Goo, J., Huang, C., Yoo, C., & Koo, C. (2022). Smart tourism technologies' ambidexterity:
Balancing tourist's worries and novelty seeking for travel satisfaction. *Information Systems Frontiers*, *24*(6), 2139–2158. https://doi.org/10.1007/s10796-021-10233-6

Gretzel, U., Σιγάλα, M., Xiang, Z., & Koo, C. (2015). Smart tourism: Foundations and developments. *Electronic Markets*, *25*(3), 179–188. https://doi.org/10.1007/s12525-015-0196-8

Gretzel, U., & Scarpino-Johns, M. (2018). Destination resilience and smart tourism destinations. Tourism Review International, 22(3-4), 263-276.

Gupta, A., Dogra, N., & George, B. (2018). What determines tourist adoption of smartphone apps? *Journal of Hospitality and Tourism Technology*, *9*(1), 50–64. https://doi.
org/10.1108/jhtt-02-2017-0013

Hamid, M., Rahmat, N., & Azmadi, A. (2023). Stakeholders perception of smart tourism
technology for tourism destination. *International Journal of Academic Research in Business and Social Sciences*, *13*(4). https://doi.org/10.6007/ijarbss/v13-i4/16624

Harrison, R., DeSilvey, C., Holtorf, C., Macdonald, S., Bartolini, N., Breithoff, E., Fred-
heim, H., Lyons, A., May, S., Morgan, J., & Penrose, S. (2020). *Heritage futures: Comparative approaches to natural and cultural heritage practices*. UCL Press.

Hofer, L., Zanini, M., & Faleschini, F. (2016). Analysis of the 2016 amatrice earthquake
macroseismic data. *Annals of Geophysics*, *59*. https://doi.org/10.4401/ag-7208

Imtiaz, H., & Suki, N. M. (2022). Mobile travel apps engagement: Measuring tourists' perception. *International Journal of Interactive Mobile Technologies (iJIM)*, *16*(14), 171–
181. https://doi.org/10.3991/ijim.v16i14.31445

Ivars-Baidal, J., Celdrán Bernabéu, M. A., & Femenia-Serra, F. (2017). Guía de Implant-
ación de Destinos Turísticos Inteligentes de la Comunitat Valenciana.

Jalilvand, M. R., & Samiei, N. (2012). The impact of electronic word of mouth
on a tourism destination choice. *Internet Research*, *22*(5), 591–612. https://doi.
org/10.1108/10662241211271563

Jeong, M., & Shin, H. H. (2019). Tourists' experiences with smart tourism technology at
smart destinations and their behavior intentions. *Journal of Travel Research*, *59*(8), 1464–
1477. https://doi.org/10.1177/0047287519883034

Kim, D., Park, J., & Morrison, A. (2008). A model of traveller acceptance of mobile technology. *International Journal of Tourism Research*, *10*(5), 393–407. https://doi.org/10.1002/
jtr.669

Kirkpatrick, L. O. (2015). Urban triage, city systems, and the remnants of community:
Some "sticky" complications in the greening of Detroit. *Journal of Urban History*, *41*(2),
261–278.

Lan, F., Huang, Q., Zeng, L., Guan, X., Xing, D., & Cheng, Z. (2021). Tourism experience
and construction of personalized smart tourism program under tourist psychology. *Frontiers in Psychology*, *12*, 691183. https://doi.org/10.3389/fpsyg.2021.691183

Lee, P., Hunter, W. C., & Chung, N. (2020). Smart tourism city: Developments and transformations. *Sustainability*, *12*(10), 3958. https://doi.org/10.3390/su12103958

Liang, F., Mu, L., Wang, D., & Kim, B. (2021). A new model path for the development of smart leisure sports tourism industry based on 5g technology. *IET Communications*, *16*(5), 485–496. https://doi.org/10.1049/cmu2.12271

Lin, S., Juan, P., & Lin, S. (2020). A tam framework to evaluate the effect of smartphone application on tourism information search behavior of foreign independent travelers. *Sustainability*, *12*(22), 9366. https://doi.org/10.3390/su12229366

Liu, D., Wu, L., & Li, R. (2018). Social media envy: How experience sharing on social networking sites drives millennials' aspirational tourism consumption. *Journal of Travel Research*, *58*(3), 355–369. https://doi.org/10.1177/0047287518761615

Marra, A., Trizio, I., & Fabbrocino, G. (2021). Digital tools for the knowledge and safeguard of historical heritage. In *International Workshop on Civil Structural Health Monitoring* (pp. 645–662). Springer International Publishing.

Molinillo, S., Anaya-Sánchez, R., Morrison, A., & Coca-Stefaniak, J. (2019). Smart city communication via social media: Analysing residents' and visitors' engagement. *Cities*, *94*, 247–255. https://doi.org/10.1016/j.cities.2019.06.003

Müller, T., Schuberth, F., & Henseler, J. (2018). PLS path modeling – a confirmatory approach to study tourism technology and tourist behavior. *Journal of Hospitality and Tourism Technology*, *9*(3), 249–266. https://doi.org/10.1108/JHTT-09-2017-0106

Neuhofer, B., Buhalis, D., & Ladkin, A. (2015). Smart technologies for personalized experiences: A case study in the hospitality domain. *Electronic Markets*, *25*(3), 243–254. https://doi.org/10.1007/s12525-015-0182-1

Papadimitriou, D., Apostolopoulou, A., & Kaplanidou, K. (2013). Destination personality, affective image, and behavioral intentions in domestic urban tourism. *Journal of Travel Research*, *54*(3), 302–315. https://doi.org/10.1177/0047287513516389

Pencarelli, T. (2019). The digital revolution in the travel and tourism industry. *Information Technology & Tourism*, *22*(3), 455–476. https://doi.org/10.1007/s40558-019-00160-3

Porto, F., Silva, B., Costa, C., & Modena, C. (2012). Macro-scale analysis of damage to churches after earthquake in Abruzzo (Italy) on April 6, 2009. *Journal of Earthquake Engineering*, *16*(6), 739–758. https://doi.org/10.1080/13632469.2012.685207

Qiu, Q., Zuo, Y., & Zhang, M. (2022). Intangible cultural heritage in tourism: Research review and investigation of future agenda. *Land*, *11*(1), 139.

Santos-Júnior, A., García, F., Morgado, P., & Filho, L. (2020). Residents' quality of life in smart tourism destinations: A theoretical approach. *Sustainability*, *12*(20), 8445. https://doi.org/10.3390/su12208445

Sun, W., Tang, S., & Liu, F. (2021). Examining perceived and projected destination image: A social media content analysis. *Sustainability*, *13*(6), 3354.

Sustacha, I., Baños-Pino, J. F., & Del Valle, E. (2023). The role of technology in enhancing the tourism experience in smart destinations: A meta-analysis. *Journal of Destination Marketing & Management*, *30*, 100817. https://doi.org/10.1016/j.jdmm.2022.100817

World Travel Tech Awards. (2022). World's Best Smart Tourism Destination. https://worldtraveltechawards.com/award/world-best-smart-tourism-destination/2022

Xiang, Z., & Gretzel, U. (2010). Role of social media in online travel information search. *Tourism Management*, *31*(2), 179–188.

Xu, F., Huang, S., & Li, S. (2019). Time, money, or convenience: What determines Chinese consumers' continuance usage intention and behavior of using tourism mobile apps? *International Journal of Culture Tourism and Hospitality Research*, *13*(3), 288–302. https://doi.org/10.1108/ijcthr-04-2018-0052

Yang, J., & Wang, L. (2021). Augmented reality in cultural heritage: A review. *Journal of Cultural Heritage*, *52*, 102–113.

Ye, S., Li, X., & Wang, D. (2021). Stakeholder engagement in smart tourism destination development: A case study of Mainland China and Hong Kong. *Journal of Destination Marketing & Management*, *19*, 100517. https://doi.org/10.1016/j.jdmm.2020.100517

Yin, F., Yin, X., Zhou, J., Zhang, X., Zhang, R., Ibeke, E., Iwendi, M. G., & Shah, M. (2022). Tourism cloud management system: The impact of smart tourism. *Journal of Cloud Computing: advances, systems and applications* [online], *11*(1), 37. https://urlsand.esvalabs.com/?u=https%3A%2F%2Fdoi.org%2F10.1186%2Fs13677-022-00316-3&e=44afb75b&h=0f72853f&f=y&p=y

Zeng, Z., Chen, P., Xiao, X., Liu, P., & Zhang, J. (2022). The mediating and moderating effects on the intention to use navigation apps. *Journal of Hospitality and Tourism Technology*, *13*(5), 972–991. https://doi.org/10.1108/jhtt-07-2021-0200

Zhou, X., Chengcai, T., Lv, X., & Xing, B. (2020). Visitor engagement, relationship quality, and environmentally responsible behavior. *International Journal of Environmental Research and Public Health*, *17*(4), 1151. https://doi.org/10.3390/ijerph17041151

Zvaigzne, A., Mietule, I., Kotane, I., Vonoga, A., & Meiste, R. (2023). *Smart tourism: The role and synergies of stakeholders*. Worldwide Hospitality and Tourism Themes.

13 Impact of the Metaverse on Tourism and Hospitality Industry

Suneel Kumar, Varinder Kumar, Nisha Devi, and Isha Kumari Bhatt

Introduction

In today's rapidly evolving digital landscape, both in India and across the globe, the digitization of various industries is paving the way for unprecedented advancements in travel, tourism, and aviation. The impact of this transformation is particularly pronounced with the advent of 5G technology, which promises to further revolutionize these sectors (Monaco, 2021). In the midst of this digital revolution, information and communication technologies have played a pivotal role in reshaping the hospitality and tourism industries (Jayawardena et al., 2023; Bethapudi, 2013). These technologies have equipped hospitality and tourism businesses with strategic tools to enhance customer experiences through interactive and intelligent solutions (Buhalis et al., 2019; Neuhofer et al., 2015).

Yet, digital transformation is not confined to conventional technologies; it extends into the realm of the metaverse. The metaverse represents the next frontier of disruptive technology, one poised to exert a profound and lasting influence on society in the coming decades (Lee et al., 2021; Yawised et al., 2022). Although it is still in its conceptual stage, the metaverse holds the promise of delivering immersive experiences that seamlessly blend the virtual and real worlds (Venugopal et al., 2023). It enables users to traverse between these two realms, offering a novel and engaging user experience.

The term "Metaverse" denotes more than a fleeting trend; it signifies a fundamental shift in the way individuals interact with technology. This shift is made possible by the convergence of a myriad of cutting-edge technologies, including the Internet of Things, artificial intelligence, blockchain, and more (Allam et al., 2022; Terry & Keeney, 2022; Kumar et al., 2021). In essence, the metaverse brings together social networking platforms, online gaming environments, AR, VR, and cryptocurrency into a cohesive digital universe (Ning et al., 2023). Augmented reality, as a core component of the metaverse, enhances the user experience by superimposing digital images, sounds, and sensory inputs onto the physical world (Abrash, 2021; Gupta et al., 2022; Valeri, 2024).

The metaverse delivers a continuous, 24×7 experience, offering users the ability to virtually interact with one another and explore a boundless digital universe. While it does not adhere to a specific technology, it represents a paradigm shift

DOI: 10.4324/9781003497004-15

in human engagement with technology, bridging the gap between reality and the digital realm (Karagoz Zeren, 2023; Terry & Keeney, 2022).

Early pioneers in this virtual frontier have already begun constructing digital realms, functioning as virtual reality playgrounds and workspaces that establish novel channels of engagement with customers (Messinger et al., 2009). Take, for instance, the case of M Social and Millennium Hotels in Singapore, which have ventured into the metaverse (Chicotsky, 2023). Millennium Hotels & Resorts made history by becoming the first hospitality company to operate a hotel in the Metaverse, introducing M Social Decentraland in 2013 (Chen et al., 2023). This virtual hotel, modeled after M Social hotels worldwide, aimed to serve as a brand-building initiative with the objectives of attracting new consumers, creating immersive experiences, engaging with customers, and establishing a modern brand identity. The outcome was resoundingly successful, generating global PR value estimated at around $500,000 and leaving a lasting impact on brand perception.

As the world transitions into the era of Industry 4.0, the travel and tourism sector is compelled to embrace the transformative potential of technology. The outbreak of the pandemic has accelerated the urgency of leveraging digital advancements (Haywood, 2020; Lew et al., 2020). Recent research conducted by Google underscores the increasing reliance on the Internet for travel planning, with 74% of holidaymakers turning to online resources (Law et al., 2020). Additionally, the proliferation of mobile devices has prompted the travel and tourism industry to prioritize the development of travel applications to enhance the travel experience. Remarkably, 85% of travelers now plan their holiday trips using travel apps, indicating a significant shift in consumer behavior (Buhalis & Foerste, 2015; Sebby et al., 2022).

This study is essential due to the evolving nature of technology and its profound impact on the tourism and hospitality sectors. With the emergence of the metaverse, there is a critical need to understand how this transformative technology can reshape these industries. The study addresses the research gap by exploring the uncharted territory of the metaverse's implications, its benefits, challenges, and its potential for enhancing customer experiences and business operations in tourism and hospitality. By conducting this research, we can gain valuable insights into how these sectors can adapt and thrive in a rapidly changing digital landscape.

Metaverse in Tourism and Hospitality

The COVID-19 pandemic and geopolitical conflicts have once again highlighted the vulnerability of the hotel and tourism industry to external and internal disruptions. Simultaneously, the growing awareness of climate change is expected to influence travelers' decisions regarding hotel stays and travel destinations (González-Reverté et al., 2022; Seyfi et al., 2023). These shifting consumer preferences, coupled with generational transitions and other external factors, may drive a shift toward supporting sustainable hospitality establishments and favoring local travel experiences over international ones (Dwivedi et al., 2022, 2023).

The evolving landscape of customer behavior underscores the importance of the hospitality and tourism sectors' readiness to design and provide authentic experiences within the metaverse. As the required technology matures and user acceptance increases, metaverse applications are poised to play a significant role in these sectors, enhancing marketing, customer interactions, communication, decision-making processes, and overall visitor experiences (Batat & Hammedi, 2023; Wong et al., 2023). Consequently, metaverse experiences in the realm of hospitality and tourism could potentially become the next disruptive force in these industries (Yang & Wang, 2023).

The integration of the metaverse into the tourism and hospitality sector yields significant benefits with profound academic and practical implications. By offering immersive and interactive experiences, metaverse tourism inspires travelers and fosters a deep emotional connection, translating into heightened enthusiasm for travel (Dwivedi et al., 2022; Gursoy et al., 2023). Furthermore, metaverse solutions, particularly VR and AR technologies, provide detailed and realistic insights, empowering potential customers with a comprehensive understanding of accommodations and destinations (Buhalis, Leung, et al., 2023; Koohang et al., 2023; Kouroupi & Metaxas, 2023). This reduction in uncertainty, in turn, enhances traveler confidence, thereby increasing the likelihood of successful bookings. Metaverse technology not only bolsters customer engagement and brand loyalty but also positions early adopters as industry innovators with a competitive edge (Bilgihan & Ricci, 2023; Valeri, 2023). Cross-selling opportunities and revenue maximization are realized through immersive experiences, while cost reduction and global market expansion occur through the replacement of traditional marketing methods with virtual alternatives. The metaverse also generates invaluable data for data-driven insights, facilitating continuous improvement in marketing strategies and overall customer experiences (Bibri, 2022; Ghazouani et al., 2023). In essence, the metaverse has the potential to revolutionize the tourism and hospitality sector by fundamentally altering the way travelers engage with destinations, accommodations, and services, ultimately driving bookings and revenue.

The metaverse is expected to have a profound impact on hospitality and tourism, revolutionizing the guest experience across various phases, including before, during, and after travel. The breadth of services within the hospitality sector, encompassing lodging, food and beverage, and entertainment, presents both significant opportunities and challenges when blending the physical and virtual realms. This transformation will require a reengineering of business processes, affecting operational and strategic aspects of hospitality management (Çolakoğlu et al., 2023; Gursoy et al., 2022; Wong et al., 2023). Table 13.1 illustrates the diverse array of metaverse hospitality services spanning the stages before, during, and after a physical visit (Buhalis, Leung, et al., 2023). These services range from virtually experiencing hospitality when travel is impractical to utilizing augmented reality for on-site interpretation. The metaverse offers a comprehensive framework for enhancing the overall visitor experience, encompassing ambience, information, and engagement.

Table 13.1 Metaverse hospitality services before, during, and after a physical visit

Phase of visit	Metaverse hospitality services
BEFORE physical visit	– Experience hospitality services virtually when real travel is not possible. – Enjoy ambiance and atmosphere through realistic gamification. – Conduct information collection and fact-finding before visiting. – Assess the suitability and fitness of services to meet individual needs and requirements.
DURING physical visit	– Plan daily itineraries. – Explore local resources for visiting. – Utilize AR for interpretation and understanding. – Engage with additional knowledge sources to appreciate local history or nature.
AFTER physical visit	- Re-experience places and recharge their memories. – Re-engage with people, places, resources, and cultures. – Reconnect with service providers and fellow customers. – Demonstrate destinations and organizations and share them with others. – Plan future trips and explorations.

Source: Adapted from Buhalis, Lin, et al. (2023)

As public awareness of the metaverse grows, airports and airlines are increasingly leveraging augmented and virtual reality to optimize their operations and prevent costly disruptions. SITA, a leading international IT and telecommunications provider for aviation communities, has partnered with major airports across the USA, Europe, the Middle East, and Africa to implement their Digital Twin solution (Kumar et al., 2023a; Lee et al., 2021). This virtual representation of airports allows stakeholders to assess operational status in real time, identify issues, and proactively address them, enhancing the overall airport experience.

Furthermore, businesses and brands are investing in engaging AR experiences designed to enhance customer engagement and foster brand loyalty (Jessen et al., 2020). SITA's 2022 IT Insights study revealed that 35% of airlines have long-term ambitions related to the Metaverse, although only 1% have confirmed significant initiatives. Similarly, 23% of airports have long-term plans, with 9% already implementing substantial Metaverse-related projects (SITA, 2022).

Innovative Applications of Metaverse Technology

The tourism industry is undergoing a revolutionary transformation through metaverse technology, offering travelers a diverse range of immersive experiences. In the realm of VR Tourism, virtual reality is redefining travel, enabling users to explore destinations, attend virtual concerts, and engage in entertainment events within the metaverse (Dwivedi et al., 2022; Yang & Wang, 2023). Similarly, AR enhances real-world surroundings with interactive elements, providing travelers with location-specific information and experiences (Lim et al., 2023; Yew & Chandrashekar, 2023). Online Trade Shows and Expos transition into the metaverse, combining online accessibility with the dynamics of in-person events through avatars and virtual spaces (Far et al., 2023; Sonnen, 2022). The metaverse paves the way for Virtual Theme Parks and Attractions, allowing for the creation of imaginative environments free from physical constraints.

Several pioneering businesses are leading this transformation, such as New Frontier's digital recreation of The Great Tapestry of Scotland, Emirates' 360-degree aircraft cabin tours, and Vueling's sustainable virtual travel options. The Seoul Metropolitan Government introduces "Metaverse Seoul," offering essential services in a discrimination-free environment. Rendezverse streamlines event planning, Mövenpick Hotel Amsterdam City Centre enhances venue inspection, and Qatar Airways offers interactive virtual aircraft tours. Pinktada introduces tokenization in the travel industry, the Barbados Government pioneers digital diplomacy, and Millennium Hotels operates a virtual hotel for promotional and branding purposes. These examples showcase the metaverse's dynamic impact, reimagining how travelers engage with destinations, accommodations, and attractions, and heralding a new era of tourism innovation.

Co-Creation in Tourism and Hospitality Using Metaverse

The introduction of the metaverse into the tourism and hospitality industry has transformative implications for the dynamics between customers and service providers. Through the metaverse, hospitality enterprises can seamlessly engage with customers across virtual and physical realms (Buhalis, Leung, et al., 2023; Dwivedi et al., 2022). By enabling business-to-consumer interactions, the hospitality sector can empower customers to actively co-create their experiences using technology (Neuhofer et al., 2015). This immersive approach not only enhances how the industry presents itself to consumers but also fosters co-creation, enriching the overall experience.

The ethereal nature of the metaverse enables a personalized and co-created experience that closely mirrors physical settings (Moro-Visconti, 2022). Customers can actively participate in shaping their experiences by choosing from a variety of hospitality providers, including restaurants, event venues, and leisure facilities (Buhalis, Lin, et al., 2023). As a result, hotel companies can leverage the metaverse to allow customers to preview their offerings before visiting in person, creating multiple touchpoints along the customer journey (Liu et al., 2023; Solakis et al., 2022). Furthermore, the metaverse provides a unique space for simulating potentially challenging real-world scenarios. These interactions can also extend to customers who have already experienced a service and are eager to provide feedback and suggestions (Dincelli & Yayla, 2022; Han et al., 2022). Incorporating

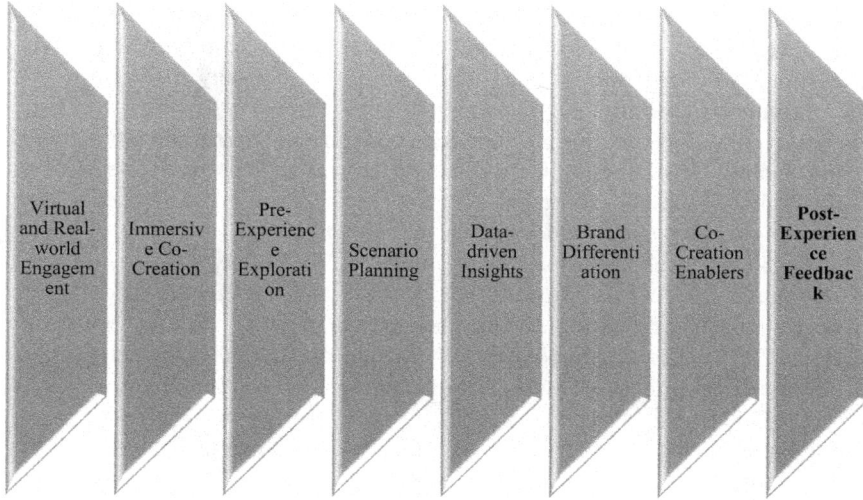

Figure 13.1 Potential of metaverse.

the metaverse into the tourism and hospitality sector not only augments customer experiences but also offers an innovative and creative approach to service delivery, thereby redefining the way businesses and customers interact and co-create in the digital age.

The integration of the metaverse in the hospitality industry presents a paradigm shift in the way customers and service providers interact. This transformative technology allows for a range of creative and innovative approaches to co-creation within the sector (Figure 13.1).

4.1 Virtual- and Real-World Engagement
The metaverse blurs the lines between the virtual and physical worlds. It enables tourism hospitality firms to engage with customers seamlessly in both realms. Customers can explore and experience services virtually, allowing them to make informed decisions before physically visiting a location (Z. Chen, 2023; Gursoy et al., 2022).

4.2 Immersive Co-Creation
The metaverse facilitates immersive co-creation experiences. Customers can actively participate in designing their experiences by choosing from a spectrum of offerings such as dining, events, and recreational activities (Buhalis, Leung, et al., 2023; Mohanty et al., 2022). This hands-on involvement enhances personalization and customer satisfaction.

4.3 Pre-Experience Exploration
Hotel companies and other service providers can leverage the metaverse as a testing ground for customers. Users can virtually sample services and products,

providing a risk-free environment to experiment and customize their experiences (Koohang et al., 2023; Kumar & Kumar, 2022). This approach not only generates more touchpoints but also fosters customer loyalty.

4.4 Scenario Planning

The metaverse offers a unique platform for simulating real-world situations, including potentially distressing or challenging scenarios. This proactive approach allows businesses to prepare for various circumstances and respond effectively, ensuring a seamless and enjoyable customer experience.

4.5 Post-Experience Feedback

In the metaverse, interactions extend beyond pre-visit engagement. Customers who have already experienced a service can continue to engage with service providers and fellow customers. They can offer feedback, share their experiences, and provide insights for ongoing improvement.

4.6 Co-Creation Enablers

Technologies such as AR and VR are instrumental in creating immersive and interactive metaverse experiences. These tools empower customers to actively shape their encounters, ensuring that co-creation is at the forefront of service delivery.

4.7 Brand Differentiation

Embracing the metaverse for co-creation not only enhances customer experiences but also sets businesses apart from competitors. It showcases a forward-thinking approach, attracting tech-savvy customers and positioning brands as leaders in the industry.

4.8 Data-Driven Insights

The metaverse generates valuable data on customer preferences, behaviors, and interactions. This information can be harnessed to fine-tune services, tailor marketing efforts, and continuously improve customer experiences.

The metaverse's role in the hospitality sector extends beyond enhancing customer interactions; it empowers customers as co-creators of their experiences. This creative and academic exploration underscores the metaverse's potential to redefine the way hospitality businesses engage with their clientele.

Metaverse Challenges

The integration of AR and VR technologies into the metaverse presents several distinctive challenges within the context of the tourism and hospitality sector. These challenges can be categorized as follows.

Cognitive and Emotional Influence: The deployment of AR and VR technologies can exert a profound cognitive and emotional influence on users. These immersive experiences have the potential to mold users' decision-making processes, behavioral responses, and emotional states. Within the context of the tourism and hospitality sector, businesses must grapple with the ethical utilization of these technologies (Jafar & Ahmad, 2023; Tsai, 2022). Understanding how these technologies influence users and ensuring that these influences result in positive and desirable customer experiences are crucial.

5.1   Economic Barrier to Adoption

The initial financial cost of acquiring the requisite hardware for AR and VR experiences is a substantial hurdle to their widespread adoption in the tourism and hospitality sector. To facilitate their broader integration and utilization, strategies must be devised to render these technologies more financially accessible (Corne et al., 2023; Dwivedi et al., 2022). As technology advances, the anticipated reduction in costs becomes a significant point of interest.

5.2   Risks in AR

AR applications, particularly those rooted in location-based interactions, introduce a spectrum of physical and psychological risks. Users immersed in AR experiences may inadvertently neglect their physical surroundings, leading to potential accidents and safety hazards (Monaco & Sacchi, 2023). Businesses must address this issue by instituting safety precautions and fostering user awareness. Furthermore, the abundance of information and sensory inputs in AR can cause cognitive overload (Yawised et al., 2023). Striking the right balance between providing information and preventing user overwhelm presents a complex challenge.

5.3   Moral and Ethical Dilemmas

Unauthorized manipulation of AR content for biased or harmful purposes emerges as a pressing moral and ethical dilemma. The potential for AR to be misused for propagandistic or misinformation purposes demands stringent ethical guidelines and oversight to prevent such misuse (Özdemir Uçgun & Şahin, 2023).

5.4   Data Privacy and Cybersecurity Implications

AR applications routinely collect user data for various purposes, raising significant data privacy concerns and the potential for legal issues arising from data mishandling (Monaco & Sacchi, 2023). Safeguarding user information through robust cybersecurity measures and data protection mechanisms is imperative to maintain user trust and security (Liyanaarachchi et al., 2023).

5.5   Health-Related Concerns in VR

VR experiences introduce various health-related concerns, including motion-related issues such as motion sickness, nausea, and dizziness. These issues pose particular challenges in the context of the tourism and hospitality sector, where user comfort is paramount (Han et al., 2022). Additionally, extended use of VR headsets can lead to physical strain, particularly in the head and neck areas (Wadhar et al., 2023). The social and psychological impact of VR, including the potential for addiction, social isolation, and detachment from real-life experiences, warrants careful consideration. VR environments can also enable anti-social behaviors, necessitating the creation of safe and moderated spaces (Yung et al., 2021). Ensuring that VR content is sensitive to user comfort and safety is essential, as exposure to violent or disturbing content in realistic VR environments can negatively impact users.

These challenges underscore the need for a cautious and responsible approach to navigate the complexities of integrating AR and VR technologies into the metaverse. This approach is essential to ensure the ethical, safe, and enjoyable use of these technologies in the tourism and hospitality industry.

Discussion and Conclusion

The metaverse is set to become a transformative force in the tourism and hospitality sector. AR and VR technologies are poised to redefine the way businesses engage with customers, providing immersive and interactive experiences that bridge the physical and virtual realms. Early pioneers in the industry, such as M Social and Millennium Hotels, have already demonstrated the metaverse's potential to reshape brand identity and engage customers in novel ways. By offering virtual environments that mimic physical settings, businesses can empower customers to co-create their experiences, fostering personalization and deeper engagement. This not only enhances customer interactions but also creates multiple touchpoints along the customer journey, ultimately redefining the way businesses and customers interact.

However, the integration of the metaverse into the tourism and hospitality sector is not without its challenges. AR and VR technologies have the power to influence decision-making, emotions, and behaviors, requiring responsible and ethical usage (Kumar et al., 2023b). The economic barrier to adoption, particularly the initial cost of hardware, remains a challenge to widespread implementation. Health and safety concerns in AR and VR experiences, from motion-related issues to potential addiction, need to be addressed to ensure user comfort and well-being. Moreover, a lack of universal understanding and shared definitions of the metaverse can create confusion and misaligned expectations, underscoring the need for clear industry standards. In conclusion, the metaverse represents a significant paradigm shift in the tourism and hospitality sector. As businesses navigate the complexities and challenges associated with metaverse adoption, they have the opportunity to position themselves as industry innovators and provide customers with unprecedented experiences. The responsible and customer-centric integration of the metaverse will be key to realizing its full potential in revolutionizing the sector.

In this era of rapidly evolving digital transformation, the metaverse presents itself as a revolutionary and disruptive force, offering immense potential for reshaping the landscape of the tourism and hospitality sector. The advent of technologies such as AR and VR within the metaverse brings forth a wealth of opportunities to redefine the relationship between businesses and customers, ultimately enhancing the customer experience and transforming operational efficiency. The metaverse offers a unique and immersive platform for co-creation within the hospitality industry, empowering customers to actively participate in shaping their experiences. Through seamless engagement across both virtual and physical realms, customers can preview offerings, explore services, and actively contribute to the design of their experiences. This innovative approach enhances personalization, fosters loyalty, and redefines the way businesses and customers interact. By enabling businesses to adapt to the changing expectations and behaviors of tech-savvy customers, the metaverse positions early adopters as industry leaders with a competitive edge.

Nevertheless, while the metaverse presents exciting opportunities, it also brings along a set of distinctive challenges that must be addressed for responsible and ethical use. The influence of AR and VR on decision-making and emotions requires careful consideration, and the economic barriers to adoption, such as the initial costs of hardware, must be overcome for widespread implementation. Additionally, health and safety concerns in AR and VR experiences, including motion-related issues and psychological impact, demand attention to ensure user well-being. In conclusion, as businesses navigate the uncharted territory of the metaverse, they have the potential to revolutionize the tourism and hospitality sector. The responsible and ethical integration of metaverse technologies will not only enhance customer interactions but also pave the way for the industry's future, marked by innovation, creativity, and an unparalleled level of customer co-creation. By addressing the challenges and harnessing the opportunities presented by the metaverse, the sector can embrace a transformative journey that promises to redefine the way travelers engage with destinations, accommodations, and attractions, ultimately driving bookings and revenue. The metaverse is not merely a trend; it is a fundamental shift that holds the power to shape the future of the industry.

Future Research Scope and Limitation

As the metaverse continues to emerge as a transformative force in the Tourism and Hospitality sector, future research endeavors offer a promising path for scholars and industry stakeholders. First, there is a growing need to delve deeper into the dynamics of co-creation within the metaverse, exploring the various ways customers and service providers interact, in both virtual and physical realms. Understanding the evolving expectations and behaviors of users will be instrumental in shaping customer experiences.

Furthermore, the integration of AR and VR technologies in the metaverse necessitates comprehensive research on the cognitive, emotional, and ethical implications of these immersive experiences. Unveiling the persuasive and influential factors of AR and VR on decision-making and emotions, alongside the ethical considerations of their deployment, will be critical for both academia and industry.

Moreover, the reduction of economic barriers to metaverse adoption, such as the affordability of hardware, should be a focal point for future research to ensure broader implementation across the sector. Research that explores strategies for minimizing initial costs and expanding accessibility will be invaluable for businesses seeking to leverage metaverse technologies.

Finally, the health and safety aspects of metaverse technologies, particularly in the context of tourism and hospitality, demand extensive study. Investigations into mitigating motion-related issues in VR experiences and addressing potential psychological challenges will be crucial for user well-being. Future research endeavors will be instrumental in shaping a responsible, efficient, and customer-centric future for the tourism and hospitality sector within the metaverse. The complex interplay of technology, ethics, and human experiences offers an exciting landscape for scholars and industry leaders to explore and contribute to this trans-formative journey.

This study is not without its limitations. First, the fast-evolving nature of metaverse technology and its applications in the tourism and hospitality sector makes it challenging to capture the full spectrum of developments within this dynamic space. While the research offers a comprehensive overview, certain specific advancements and innovations may have emerged beyond the scope of this study. Second, the metaverse's conceptual ambiguity poses a challenge in terms of aligning the varied definitions and understandings of this technology. Researchers and industry stakeholders may have different interpretations and expectations of the metaverse, which could impact the conclusions drawn from this study. Additionally, this research primarily focuses on the positive implications and benefits of metaverse integration in the tourism and hospitality sector. The study does not extensively explore the potential drawbacks, risks, or unintended consequences, such as user addiction to immersive experiences or the ethical dilemmas of data privacy in this context. A more balanced assessment would require further research into these aspects.

References

Abrash, M. (2021). *Creating the future: Augmented reality, the next human-machine interface*. Paper presented at the 2021 IEEE International Electron Devices Meeting (IEDM). https://www.researchgate.net/publication/359127160_Creating_the_Future_Augmented_Reality_the_next_Human-Machine_Interface

Allam, Z., Sharifi, A., Bibri, S. E., Jones, D. S., & Krogstie, J. (2022). The metaverse as a virtual form of smart cities: Opportunities and challenges for environmental, economic, and social sustainability in urban futures. *Smart Cities, 5*(3), 771–801.

Batat, W., & Hammedi, W. (2023). The extended reality technology (ERT) framework for designing customer and service experiences in phygital settings: A service research agenda. *Journal of Service Management, 34*(1), 10–33. https://doi.org/10.1108/JOSM-08-2022-0289

Bethapudi, A. (2013). The role of ICT in tourism industry. *Journal of Applied Economics and Business, 1*(4), 67–79.

Bibri, S. E. (2022). The social shaping of the metaverse as an alternative to the imaginaries of data-driven smart cities: A study in science, technology, and society. *Smart Cities, 5*(3), 832–874.

Bilgihan, A., & Ricci, P. (2023). The new era of hotel marketing: Integrating cutting-edge technologies with core marketing principles. *Journal of Hospitality and Tourism Technology, 15*(1). https://doi.org/10.1108/JHTT-04-2023-0095

Buhalis, D., & Foerste, M. (2015). SoCoMo marketing for travel and tourism: Empowering co-creation of value. *Journal of Destination Marketing & Management, 4*(3), 151–161.

Buhalis, D., Harwood, T., Bogicevic, V., Viglia, G., Beldona, S., & Hofacker, C. (2019). Technological disruptions in services: Lessons from tourism and hospitality. *Journal of Service Management, 30*(4), 484–506. https://doi.org/10.1108/JOSM-12-2018-0398

Buhalis, D., Leung, D., & Lin, M. (2023). Metaverse as a disruptive technology revolutionising tourism management and marketing. *Tourism Management, 97*, 104724. https://doi.org/10.1016/j.tourman.2023.104724

Buhalis, D., Lin, M. S., & Leung, D. (2023). Metaverse as a driver for customer experience and value co-creation: Implications for hospitality and tourism management and marketing. *International Journal of Contemporary Hospitality Management, 35*(2), 701–716. https://doi.org/10.1108/IJCHM-05-2022-0631

Chen, S., Chan, I. C. C., Xu, S., Law, R., & Zhang, M. (2023). Metaverse in tourism: Drivers and hindrances from stakeholders' perspective. *Journal of Travel & Tourism Marketing, 40*(2), 169–184.

Chen, Z. (2023). Beyond reality: Examining the opportunities and challenges of cross-border integration between metaverse and hospitality industries. *Journal of Hospitality Marketing & Management, 32*(7), 967–980. https://doi.org/10.1080/19368623.2023.2222029

Chicotsky, B. (2023). Web3 and marketing: The new frontier. *Applied Marketing Analytics, 9*(2), 182–194.

Çolakoğlu, Ü., Anış, E., Esen, Ö., & Tuncay, C. S. (2023). The evaluation of tourists' virtual reality experiences in the transition process to metaverse. *Journal of Hospitality and Tourism Insights, ahead-of-print*(ahead-of-print). https://doi.org/10.1108/JHTI-09-2022-0426

Corne, A., Massot, V., & Merasli, S. (2023). The determinants of the adoption of blockchain technology in the tourism sector and metaverse perspectives. *Information Technology & Tourism*, 1–29.

Dincelli, E., & Yayla, A. (2022). Immersive virtual reality in the age of the metaverse: A hybrid-narrative review based on the technology affordance perspective. *The Journal of Strategic Information Systems, 31*(2), 101717. https://doi.org/10.1016/j.jsis.2022.101717

Dwivedi, Y. K., Hughes, L., Baabdullah, A. M., Ribeiro-Navarrete, S., Giannakis, M., Al-Debei, M. M., Cheung, C. M., Conboy, K., & Wamba, S. F. (2022). Metaverse beyond the hype: Multidisciplinary perspectives on emerging challenges, opportunities, and agenda for research, practice and policy. *International Journal of Information Management, 66*, 102542. https://doi.org/10.1016/j.ijinfomgt.2022.102542

Dwivedi, Y. K., Hughes, L., Wang, Y., Alalwan, A. A., Ahn, S. J., Balakrishnan, J., Barta, S., Belk, R., Buhalis, D., Dutot, V., & Felix, R. (2023). Metaverse marketing: How the metaverse will shape the future of consumer research and practice. *Psychology & Marketing, 40*(4), 750–776.

Far, S. B., Rad, A. I., & Asaar, M. R. (2023). Blockchain and its derived technologies shape the future generation of digital businesses: A focus on decentralized finance and the metaverse. *Data Science and Management, 6*(3), 183–197.

Ghazouani, M., Chafiq, N., Zaher, N., Azouazi, M., & Chakir, A. (2023). Merging big data with the metaverse: A proposed architecture with multiple layers. In *Influencer marketing applications within the metaverse* (pp. 263–287). IGI Global.

González-Reverté, F., Gomis-López, J. M., & Díaz-Luque, P. (2022). Reset or temporary break? Attitudinal change, risk perception and future travel intention in tourists experiencing the COVID-19 pandemic. *Journal of Tourism Futures*. https://doi.org/10.1108/JTF-03-2021-0079

Gupta, R., He, J., Ranjan, R., Gan, W.-S., Klein, F., Schneiderwind, C., Neidhardt, A., Brandenburg, K., & Välimäki, V. (2022). Augmented/mixed reality audio for hearables: Sensing, control, and rendering. *IEEE Signal Processing Magazine, 39*(3), 63–89.

Gursoy, D., Lu, L., Nunkoo, R., & Deng, D. (2023). Metaverse in services marketing: An overview and future research directions. *The Service Industries Journal, 43*(15–16), 1140–1172. https://doi.org/10.1080/02642069.2023.2252750

Gursoy, D., Malodia, S., & Dhir, A. (2022). The metaverse in the hospitality and tourism industry: An overview of current trends and future research directions. *Journal of*

Hospitality Marketing & Management, 31(5), 527–534. https://doi.org/10.1080/193686 23.2022.2072504

Han, D.-I. D., Bergs, Y., & Moorhouse, N. (2022). Virtual reality consumer experience escapes: Preparing for the metaverse. *Virtual Reality, 26*(4), 1443–1458.

Haywood, K. M. (2020). A post COVID-19 future – tourism re-imagined and re-enabled. *Tourism Geographies, 22*(3), 599–609. https://doi.org/10.1080/14616688.2020.1762120

Jafar, R. M. S., & Ahmad, W. (2023). Tourist loyalty in the metaverse: The role of immersive tourism experience and cognitive perceptions. *Tourism Review, 79*(2), 321–336. https:// doi.org/10.1108/TR-11-2022-0552

Jayawardena, C., Ahmad, A., Valeri, M., & Jaharadak, A. A. (2023). Technology acceptance antecedents in digital transformation in hospitality industry. *International Journal of Hospitality Management, 108*, 103350.

Jessen, A., Hilken, T., Chylinski, M., Mahr, D., Heller, J., Keeling, D. I., & de Ruyter, K. (2020). The playground effect: How augmented reality drives creative customer engagement. *Journal of Business Research, 116*, 85–98. https://doi.org/10.1016/j. jbusres.2020.05.002

Karagoz Zeren, S. (2023). The tourism sector in metaverse: Virtual hotel and applications. In *Metaverse: Technologies, opportunities and threats* (pp. 373–381). Springer.

Koohang, A., Nord, J., Ooi, K., Tan, G., Al-Emran, M., Aw, E., Dennis, C., Dutot, V., & Wong, L.-W. (2023). Shaping the metaverse into reality: Multidisciplinary perspectives on opportunities, challenges, and future research. *Journal of Computer Information Systems, 63*(3), 735–765. https://doi.org/10.1080/08874417.2023.2165197

Kouroupi, N., & Metaxas, T. (2023). Can the metaverse and its associated digital tools and technologies provide an opportunity for destinations to address the vulnerability of over-tourism? *Tourism and Hospitality, 4*(2), 355–373.

Kumar, S., & Kumar, V. (2022). Potential risk and dark side of digital transformation colonialism in tourism sector: A conceptual framework. In *The Proceedings of international conference 2022* (p. 248). https://www.researchgate.net/publication/359539161_The_ Proceedings_of_International_Conference_2022_Covid-19_Digital_Transformation_ and_Tourism_Resilience

Kumar, S., Kumar, V., & Attri, K. (2021). Impact of artificial intelligence and service robots in tourism and hospitality sector: Current use & future trends. *Administrative Development: A Journal of HIPA, Shimla, 8*, 59–83.

Kumar, S., Kumar, V., Bhatt, I. K., & Kumar, S. (2023a). Mapping research trends on smart tourism: A bibliometric analysis. *Digital Transformation of the Hotel Industry: Theories, Practices, and Global Challenges*, 87–109.

Kumar, S., Kumar, V., Bhatt, I. K., Kumar, S., & Attri, K. (2023b). Digital transformation in tourism sector: Trends and future perspectives from a bibliometric-content analysis. *Journal of Hospitality and Tourism Insights*. https://doi.org/10.1108/JHTI-10-2022-0472

Law, R., Leung, D., & Chan, I. C. C. (2020). Progression and development of information and communication technology research in hospitality and tourism. *International Journal of Contemporary Hospitality Management, 32*(2), 511–534. https://doi.org/10.1108/ IJCHM-07-2018-0586

Lee, L.-H., Braud, T., Zhou, P., Wang, L., Xu, D., Lin, Z., Kumar, A., Bermejo, C., & Hui, P. (2021). *All one needs to know about metaverse: A complete survey on technological singularity, virtual ecosystem, and research agenda.* arXiv Preprint. https://arxiv.org/ abs/2110.05352

Lew, A. A., Cheer, J. M., Haywood, M., Brouder, P., & Salazar, N. B. (2020). Visions of travel and tourism after the global COVID-19 transformation of 2020. *Tourism Geographies, 22*(3), 455–466. https://doi.org/10.1080/14616688.2020.1770326

Lim, W. Y. B., Xiong, Z., Niyato, D., Cao, X., Miao, C., Sun, S., & Yang, Q. (2023). Realizing the metaverse with edge intelligence: A match made in heaven. *IEEE Wireless Communications, 30*(4), 64–71. https://doi.org/10.1109/MWC.018.2100716

Liu, S., Benckendorff, P., & Mair, J. (2023). Value co-creation through technology-mediated experiences: A research agenda. *Tourism and the Experience Economy in the Digital Era*, 85–100.

Liyanaarachchi, G., Viglia, G., & Kurtaliqi, F. (2023). Privacy in hospitality: Managing biometric and biographic data with immersive technology. *International Journal of Contemporary Hospitality Management, ahead-of-print*(ahead-of-print). https://doi.org/10.1108/IJCHM-06-2023-0861

Messinger, P. R., Stroulia, E., Lyons, K., Bone, M., Niu, R. H., Smirnov, K., & Perelgut, S. (2009). Virtual worlds – past, present, and future: New directions in social computing. *Decision Support Systems, 47*(3), 204–228.

Mohanty, M. K., Mohapatra, A. K., Samanta, P. K., Agrawal, G., & Agrawal, G. (2022). Exploring metaverse: A virtual ecosystem from management perspective. *Journal of Commerce, 43*(4), 1–11.

Monaco, S. (2021). *Tourism, safety and COVID-19: Security, digitization and tourist behaviour*. Routledge.

Monaco, S., & Sacchi, G. (2023). Travelling the metaverse: Potential benefits and main challenges for tourism sectors and research applications. *Sustainability, 15*(4), 3348.

Moro-Visconti, R. (2022). Metaverse: A digital network valuation. In R. Moro-Visconti (Ed.), *The valuation of digital intangibles: Technology, marketing, and the metaverse* (pp. 515–559). Springer International Publishing.

Neuhofer, B., Buhalis, D., & Ladkin, A. (2015). Smart technologies for personalized experiences: A case study in the hospitality domain. *Electronic Markets, 25*(3), 243–254. https://doi.org/10.1007/s12525-015-0182-1

Ning, H., Wang, H., Lin, Y., Wang, W., Dhelim, S., Farha, F., Ding, J., & Daneshmand, M. (2023). A survey on the metaverse: The state-of-the-art, technologies, applications, and challenges. *IEEE Internet of Things Journal, 10*(16), 14671–14688. https://doi.org/10.1109/JIOT.2023.3278329

Özdemir Uçgun, G., & Şahin, S. Z. (2023). How does Metaverse affect the tourism industry? Current practices and future forecasts. *Current Issues in Tourism*, 1–15.

Sebby, A. G., Jordan, K., & Brewer, P. (2022). Travel decisions: The COVID-19 paradigm shift on the use of travel aggregator websites for vacation planning. *Tourism: An International Interdisciplinary Journal, 70*(2), 223–242.

Seyfi, S., Hall, C. M., & Shabani, B. (2023). COVID-19 and international travel restrictions: The geopolitics of health and tourism. *Tourism Geographies, 25*(1), 357–373. https://doi.org/10.1080/14616688.2020.1833972

SITA (Producer). (2022). *Meet the megatrends – SITA*. www.sita.aero/globalassets/docs/other/innovation/meet-the-megatrends/sita-megatrends-report.pdf

Solakis, K., Katsoni, V., Mahmoud, A. B., & Grigoriou, N. (2022). Factors affecting value co-creation through artificial intelligence in tourism: A general literature review. *Journal of Tourism Futures, 10*(1), 116–130. https://doi.org/10.1108/JTF-06-2021-0157

Sonnen, J. (2022). *Metaverse for beginners 2023: The ultimate guide on investing in metaverse, blockchain gaming, virtual lands, augmented reality, virtual reality, NFT, real estate, crypto and web 3.0*. Justin Sonnen.

Terry, Q., & Keeney, S. (2022). *The metaverse handbook: Innovating for the internet's next tectonic shift*. John Wiley & Sons.

Tsai, S.-P. (2022). Investigating metaverse marketing for travel and tourism. *Journal of Vacation Marketing*, 1–10. https://doi.org/10.1177/13567667221145715

Valeri, M. (2023). Knowledge management and knowledge sharing. *Business Strategies and an Emerging Theoretical Field*. https://doi.org/10.1007/978-3-031-37868-3

Valeri, M. (Ed.). (2024). *Innovation strategies and organizational culture in tourism: Concepts and case studies on knowledge sharing*. Taylor & Francis.

Venugopal, J. P., Subramanian, A. A. V., & Peatchimuthu, J. (2023). The realm of metaverse: A survey. *Computer Animation and Virtual Worlds, 34*, e2150.

Wadhar, S. B., Shahani, R., Zhou, R., Siddiquei, A. N., Ye, Q., & Asmi, F. (2023). *What factors will influence Chinese international traveling for leisure in the post-COVID-19 era: Role of health priorities and health-related information literacy.* Paper presented at the Healthcare. https://pubmed.ncbi.nlm.nih.gov/36766891/

Wong, L.-W., Tan, G. W.-H., Ooi, K.-B., & Dwivedi, Y. K. (2023). Metaverse in hospitality and tourism: A critical reflection. *International Journal of Contemporary Hospitality Management.* https://doi.org/10.1108/IJCHM-05-2023-0586

Yang, F. X., & Wang, Y. (2023). Rethinking metaverse tourism: A taxonomy and an agenda for future research. *Journal of Hospitality & Tourism Research.* https://doi.org/10.1177/10963480231163509

Yawised, K., Apasrawirote, D., & Boonparn, C. (2022). From traditional business shifted towards transformation: The emerging business opportunities and challenges in "metaverse" era. *INCBAA, 2022,* 162–175.

Yawised, K., Apasrawirote, D., Chatrangsan, M., & Muneesawang, P. (2023). Travelling in the digital world: Exploring the adoption of augmented reality (AR) through mobile application in hospitality business sector. *Journal of Advances in Management Research, 20*(4), 599–622.

Yew, N. K., & Chandrashekar, R. (2023). Analysis of technology foresight for metaverse in tourism sector by integrating quantitative approaches. *Journal of Numerical Optimization and Technology Management, 1*(1), 9–21.

Yung, R., Khoo-Lattimore, C., & Potter, L. E. (2021). Virtual reality and tourism marketing: Conceptualizing a framework on presence, emotion, and intention. *Current Issues in Tourism, 24*(11), 1505–1525.

Index

Note: Page numbers in *italics* indicate a figure and page numbers in **bold** indicate a table on the corresponding page.

For Product Safety Concerns and Information please contact our EU
representative GPSR@taylorandfrancis.com
Taylor & Francis Verlag GmbH, Kaufingerstraße 24, 80331 München, Germany